PUZZLE ME TWICE

Also by Alex Bellos

The Language Lover's Puzzle Book

Perilous Problems for Puzzle Lovers

Can You Solve My Problems?

Puzzle Ninja

Here's Looking at Euclid

The Grapes of Math

Futebol: The Brazilian Way of Life

with Edmund Harriss

Patterns of the Universe:
A Coloring Adventure in Math and Beauty

Visions of the Universe:
A Coloring Journey Through Math's Great Mysteries

PUZZLE ME TWICE

70 SIMPLE PUZZLES (ALMOST) EVERYONE GETS WRONG

Alex Bellos

Illustrations by Arnaud Boutin

THE EXPERIMENT

NEW YORK

Puzzle Me Twice: *70 Simple Puzzles (Almost) Everyone Gets Wrong*

Originally published in the UK as *Think Twice* by Square Peg, an imprint of Vintage, part of the Penguin Random House group of companies. First published in North America in revised form by The Experiment, LLC.

The Experiment, LLC
220 East 23rd Street, Suite 600
New York, NY 10010-4658
theexperimentpublishing.com

The Experiment's books are available at special discounts when purchased in bulk for premiums and sales promotions as well as for fund-raising or educational use. For details, contact us at info@theexperimentpublishing.com.

Library of Congress Cataloging-in-Publication Data available upon request

ISBN 979-8-89303-028-0
Ebook ISBN 979-8-89303-029-7

Cover design by Dan Mogford
Text design by Dan Prescott at Couper Street Type Co.
Additional text design by Beth Bugler

Manufactured in the United States of America

First printing October 2024
10 9 8 7 6 5 4 3 2 1

To the children and teachers
of Rosa Parks Elementary School, Berkeley

Contents

Introduction

I hope you get the answers wrong to every puzzle in this book.

Reader, don't let me down. I'm being honest. The more questions you answer incorrectly, the more fun you will have.

In the following pages are 70 of my favorite puzzles for which people invariably come to the wrong conclusions. When I first encountered these problems, I too fell into the traps they set. My intuitions were shaky, my lines of thought easily misdirected, my logical sharpness embarrassingly blunt.

Yet when I saw the answers, each time the "aha!" moment brought a giddy joy. The absurdity of my wrong thinking was such that I slapped my forehead with a beaming smile. These brainteasers are the most pleasurable and life-affirming type of puzzle. The ease with which they reliably provoke an erroneous answer sparks a wonder about the way in which we humans tend to think, reason, and perceive, and also a wonder about the world itself.

I've chosen *simple* puzzles, that is, questions that are succinct, easy to understand, and for which there is a single clear answer—usually yes/no, or one of a multiple choice. And I have chosen puzzles that tend to stump people at first, that is, questions and problems with counterintuitive or confounding answers and solutions. These puzzles are not ones for which you *can't* figure out an answer, they are ones where you quite quickly come to an answer, and that answer is wrong. Despite their simplicity, however, many of these puzzles touch on deep ideas.

In some cases, the claim that most people answer incorrectly is backed by data. In 2016, for example, I set the puzzle "Wandering Eyes" (page 13) in my online puzzle column for the *Guardian* newspaper, under the headline: "The logic question almost everyone gets wrong." More than 200,000 readers submitted an answer, and more than 72 percent proved the headline to be true. In other words, I warned a self-selecting audience of puzzle lovers that they would make a mistake, and most of them did.

That puzzle was my most-viewed column until 2023, when I set "Card Sharp" (page 93) with the almost identical headline: "The simple puzzle almost everyone gets wrong." It notched up millions of views, and was the *Guardian*'s most-viewed, most-shared, and most-commented article that day. People, I realized, cannot resist a challenge to their intelligence.

Now you have been warned that you will fail the tests in this book, is there anything more likely to make you want to prove me wrong?

I have sourced these conundrums from scientific journals, books, my own experience as a puzzle setter and lecturer, and my circle of teachers, magicians, and friends. I've taken problems from psychology, mathematics, statistics, physics, geography, the science of perception, and more. I've avoided trick questions, such as puzzles that rely on ambiguous phrasing and thus are designed to make you feel foolish. Instead, I have chosen puzzles that celebrate seemingly paradoxical or unexpected results. That pit gut instinct against measured reflection.

That toy with our inbuilt biases and the many assumptions we make about our environment. That catch us off guard. The best puzzles always have an element of surprise.

To get these posers right the first time you will have to overrule your intuition. Think once and you will stumble. Think twice and you are in with a chance. Given that you know I want you to fall into the traps I have set, however, the problems provide a layered challenge. Not only am I asking you to solve a puzzle, but I am asking you to solve a puzzle *about* a puzzle. The meta-problem is to unpick the deception, to understand why the solution that *feels* correct is most reliably not. What is the right answer? What is the answer that most people give? And just *why* are we fooled? These are problems that lend themselves to be solved in a group, shared, discussed, and chewed upon.

The problems are in random order. If I had grouped them by subject, you would begin to know what to look out for. The experience would be less entertaining. Enjoy this book as you would a box of chocolates: You will not know in advance what flavor of problem you are going to encounter, but you will find them moreish, each one whetting your appetite for further problem-solving and debate. Likewise, the level of trickiness varies throughout the book. My aim is to keep you all on your toes.

The answers to each question are presented immediately as you turn the page, on the left-hand side, along with some thoughts on the kind of mistakes we consistently make. Puzzles, of course, provide fleeting moments of fun. But they are also brain-sharpening tools. They provide a safe space to be duped, giving us the smarts not to be hoodwinked in the future. Once we learn the mistake, and why we made it, we are unlikely to make it again. By playfully reminding us of our own fallibility, the puzzles in this book teach us how to think more clearly.

Embrace the misdirection. Delight in the deception.

I have faith in you to stumble at every hurdle—I know you can do it if you really put your minds to it!

Silly Sum

In this puzzle you are going to be adding up numbers in your head.

Start with a thousand.
Add forty.
Add another thousand.
Add thirty.
Add another thousand.
Add twenty.
Add another thousand.
Now add ten.

What's the answer?

Did you think the answer was 5,000?

Most people do, and they are wrong. If you were writing this sum down, however, you probably wouldn't make the same mistake:

$$
\begin{array}{r}
1000 \\
40 \\
1000 \\
30 \\
1000 \\
20 \\
1000 \\
+\quad 10 \\
\hline
4100
\end{array}
$$

When you do the sum quickly in your head, it is easy to muddle the hundreds and thousands column. Adding 10 to 90 makes 100, of course, so the answer is 4,100 rather than 5,000.

Teapot Trouble

Which teapot contains more tea when full?

The shorter teapot on the left holds more tea when the pots are full because it has a higher spout. The level of tea in a full pot is the level of the spout, and can never exceed this level, because if it did, the tea would overflow and leak out.

However, if you somehow blocked the spouts, then the taller teapot would hold more tea, since you would be able to fill it up to its full volume.

To take advantage of a high teapot with a low, unblocked spout, you might think of installing an internal chamber with a high mouth by the spout, as below. The tea in the main section of the pot now will fill up to the height of the chamber.

Pint-Sized Problem

Which is longer: the height of a pint glass, or the circumference of its rim?

Surprisingly, the circumference is longer, and by quite a way. For a traditional pint glass—the one with the bulge just below the top rim, known as the "nonic pint glass" and illustrated on the previous page—the circumference is almost double the height!

The rim of a traditional British pint glass has a diameter of just under 9 cm, which gives it a circumference of just over 27 cm. (The circumference is pi, or 3.14 to two decimal places, times the diameter.) The height of a traditional pint glass is only about 15 cm.

Even the higher style of pint glasses, such as the ones often used for pilsners, have longer circumferences. The tall Peroni pint glass, for example, has a diameter of 8 cm, giving it a circumference of just over 25 cm, compared with a height of 24 cm.

Moral: We are bad at intuiting circumferences. We are used to measuring straight lines, so when we look at a glass side-on we see the circumference as from left to right and back again. But the circumference is not double the diameter—it is more than three times as long.

Wandering Eyes

Jack is looking at Anne, but Anne is looking at George. Jack is married, but George is not.

Is a married person looking at an unmarried person?

a) Yes
b) No
c) Cannot be determined

a) Yes. A married person is looking at an unmarried person.

When I set this puzzle in my *Guardian* column in 2016, 200,000 people responded, and only 28 percent of them got the right answer. That was actually a pretty good result, since it is said that only 20 percent of people get it right—by far the most common response is c).

I knew my readers were a smart bunch!

This is the explanation: We don't know the marital status of Anne. But she is either married or unmarried. If she is married, then a married person (Anne) is looking at an unmarried one (George). If she is not married, then a married person (Jack) is looking at an unmarried one (Anne). So, whatever Anne's marital status, a married person is looking at an unmarried one.

This puzzle was written by Hector Levesque, professor emeritus at the department of computer science at the University of Toronto, and is much discussed in psychology studies. When presented with only two options—a) Yes, or b) No—people are forced to think about Anne's marital status, and will probably get the correct answer.

But when there is a third option, suggesting that the answer is c) Cannot be determined, our brains take the easy way out. We are so easily nudged by the possibility that we don't have enough information to solve the question that we refrain from doing any effortful reasoning.

Why then are we so suggestible? It is because trying to reason from an "unknown quantity," in this case Anne's marital status, is cognitively hard. The chance of not having to think is just too tempting. It gives us a free pass.

The moral of the story is that *even if* a quantity is unknown, we may still be able to use it to deduce provable facts. It just takes a moment of reflection, and for us to think again.

Here's an analogous situation in mathematics, suggested by the popular math YouTuber James Grime. It requires a small bit of technical knowledge: that there are two types of number, rational ones and irrational ones.

Rational numbers have decimal expansions with a finite number of nonzero digits, or that repeat in loops forever, such as 2, 4.56, or 0.3333 . . . where the 3s repeat ad infinitum.

On the other hand, the decimal expansion of an irrational number never ends, and never repeats in a loop: Famous irrational numbers include pi and $\sqrt{2}$.

Numbers are either rational or irrational, in the same way that people are either married or not.

Right. The question James Grime asked was: Can an irrational number to the power of an irrational number ever be a rational number?

The answer is yes, and we can deduce this in exactly the same way as we worked out that yes, a married person *is* looking at an unmarried one.

We're going to use the $\sqrt{2}$ for our proof. The number $\sqrt{2}$ is irrational: It begins 1.41421 . . . and continues forever without repeating.

Let $\sqrt{2}^{\sqrt{2}} = x$. [statement A]

And let's take x to the power of $\sqrt{2}$.

$x^{\sqrt{2}} = (\sqrt{2}^{\sqrt{2}})^{\sqrt{2}} = \sqrt{2}^{\sqrt{2} \times \sqrt{2}} = (\sqrt{2})^2 = 2$. [statement B]

We know x (that is, $\sqrt{2}^{\sqrt{2}}$) is either rational or irrational. If x is rational, then according to statement A, an irrational number ($\sqrt{2}$) to the power of an irrational number ($\sqrt{2}$) is rational.

If x is irrational, then according to statement B, an irrational number (x) to the power of an irrational number ($\sqrt{2}$) is rational, since 2 is a rational number.

In other words, just as we could deduce whether a married person is looking at an unmarried one without knowing whether Anne is married, we can deduce whether an irrational number to the power of an irrational number is rational without knowing whether x is rational.

In both cases—x being rational and x being irrational—we get an example of an irrational number to the power of an irrational number being rational.

The numerical status of x, like the marital status of Anne, is irrelevant.

Confounding Cash

A woman steals $100 from a shop's cash register.

She then buys $70 of goods from that store, and gets $30 change.

How much money did the shop lose?

The shop lost $100 because that is how much the woman stole from it.

The error people make in answering this question is overthinking it. You can tie yourself in knots working out what to add and subtract.

The easiest way to think about it is to imagine two different people walking into the shop on that day:

> One of them steals $100 from the cash register.
> The other comes into the shop with $100 in her wallet, spends $70, and gets $30 change.

The actions of the second person result in the shop losing nothing. Indeed, the shop has made a sale, and the exchange of money for stuff is precisely what shops do. When a shop sells you something, the shop is not "losing" anything. It is selling it!

The only money the shop loses is the money stolen from it.

Cosmic Cranium-Cruncher

Which planet is closest, on average, to Earth?

a) Mercury
b) Venus
c) Mars
d) Jupiter

a) Mercury.

The "obvious" answer is Venus, since when we list the planets we usually order them in terms of distance from the Sun, in other words: Mercury, Venus, Earth, Mars, Jupiter, Saturn, Uranus, and Neptune.

The orbit of Venus is closest to the orbit of Earth, so Venus is the planet that travels closest.

Yet the question did not ask which planet travels *closest* to the Earth. It asked about the closest *on average*. For about half of its orbit, Venus is on the opposite side of the Sun from the Earth. Mercury is therefore, *on average*, closest to the Earth, since when Mercury is on the opposite side of the Sun it is less far away from the Earth than Venus is. For the same reason, Mercury is the closest planet, on average, not just to Earth but to *all* the planets in the solar system.

Mind Blown

Two lit candles are standing side by side, as shown below. If you take a straw and blow between them, in which direction do the flames move?

a) Away from you
b) Away from each other
c) Toward each other

c) The candle flames move toward each other.

When you blow out through the straw, your breath creates a corridor of low pressure between the candles, which draws in the air (and hence the flames) from the sides.

The Hole Truth

A flat metal plate has a circular hole cut in it. The plate is evenly heated up. When metal heats, it expands. What happens to the hole?

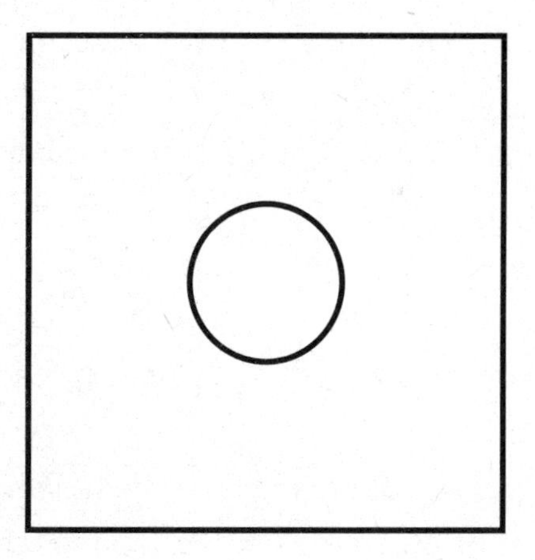

a) It gets bigger.
b) It stays the same size.
c) It gets smaller.

a) It gets bigger.

When you heat a metal, it expands because the molecules in the metal are moving further apart. (This is a useful fact to remember when you can't get the metal lid off a jam jar—run the jar lid under the hot tap for a few minutes. The metal in the lid will expand, making it easier to twist open.)

When you heat a flat metal plate, its overall dimensions will get bigger. Now what happens if the plate has a hole in it? The intuitive answer, perhaps, is that the metal will expand *into* the hole as well as expanding outward. Yet, this is not the case. If the plate is heated evenly, the average distance between the molecules grows uniformly everywhere, which means that the hole has to expand too. (If the metal expanded into the hole, some of the molecules would have to be pushed together.)

One way to think about it is to imagine the plate divided into a 3 × 3 grid, with the circle in the middle square—like having an O in the central square of a game of tic-tac-toe. When the plate heats up, each of the eight squares around the outside expand equally, which means that the dimensions of the central square (containing the hole) must also expand by the same amount.

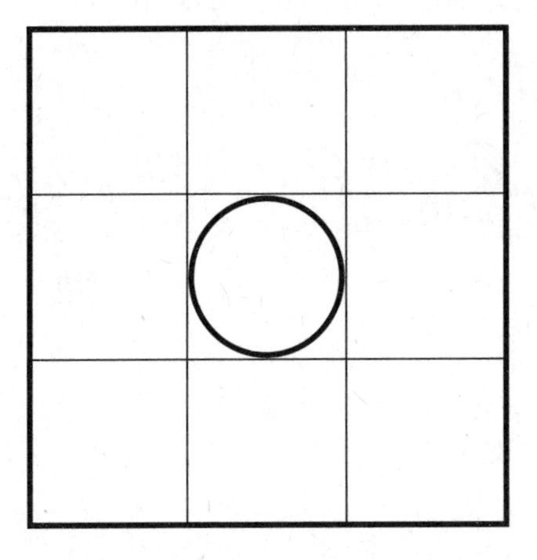

Sistery Mystery

Imagine a country where there are slightly more girls than boys. Let's say the ratio is 51 to 49—so for every 100 randomly chosen people who are male or female, on average 51 are female and 49 are male.

Who has more sisters, on average?

a) Girls
b) Boys

(In other words, on average, does each girl have more or fewer sisters than each boy?)

The answer is neither.

In this imaginary country, there are marginally more girls than boys, and this excess of girls leads to the intuitive (wrong) answer that girls probably have more sisters than boys. There are more girls around, so more all-girl families—think of all those sisters!

In fact, however, the girls and boys have, on average, *exactly the same number of sisters*. The chance of any person having a sister is more likely than the chance they have a brother, but in any family these chances are equal for both boys and girls.

To see why, let's simplify things by considering a country where the boy/girl ratio is 50/50. In two-child families, the four equally likely scenarios of children, listed in order of birth, are:

BB, BG, GB, GG

Consider the boys. Two of the boys have a sister (BG, GB), and two of the boys don't (BB).

Likewise, two of the girls have a sister (GG), and two don't (BG, GB).

Therefore, the chance of any boy or girl having a sister is 50 percent.

Now let's consider a three-child family. The eight equally likely scenarios are:

BBB, BBG, BGB, GBB, BGG, GBG, GGB, and GGG

Again, consider the boys. Nine boys have a sister (BBG, BGB, GBB, BGG, GBG, and GGB), and three don't (BBB). The same goes for the girls. (BBG, BGB, and GBB don't. The rest do.) Again, for either boys or girls, the chance of having a sister is the same, which this time is 75 percent.

This pattern will continue for larger and larger families.

Now let's do the same analysis when the ratio of girls to boys is not 50/50. You will see that the chance of having a sister is also the same for boys and girls. For brevity's sake, I'm not going to choose 51/49, because that will require too much ink. Instead, let's say the ratio is 75/25, or three to one. There are now four equally likely one-child families: G, G, G, B.

And there are 16 equally likely two-child families:

GG, GG, GG, GB
GG, GG, GG, GB
GG, GG, GG, GB
BG, BG, BG, BB

(For each family where the firstborn is a G, G, G, or B, there are four equally likely second-borns.)

In this group there are 24 girls, and 18 of them have sisters, so the chance of a girl having a sister is 18/24 = 3/4.

There are eight boys, and six of them have sisters, so the chance of a boy having a sister is 6/8 = 3/4. So when the ratio of girls to boys is 75/25, the chance of having a sister is the same for both girls and boys. Using the same technique, we can show that whatever the ratio of girls to boys, both girls and boys, on average, have the same number of sisters.

Befuddling Ballot

In an election, voters were asked to list three candidates in order of preference.

Two thirds of voters preferred Anya to Babu, and two thirds of voters preferred Babu to Cristos.

Did most voters prefer Anya to Cristos?

It seems very likely that if A did better than B, and B did better than C, then A must have done better than C. And in the real world, we might assume that this is the case.

Yet we cannot draw this conclusion. Voting can behave like a game of rock paper scissors, in which each of the three candidates beats one rival and loses to the other.

It is possible that more people preferred Cristos to Anya.

Imagine, for example, that the votes broke down as follows. Two thirds preferred A to B, and two thirds preferred B to C. Yet two thirds preferred C to A.

	⅓ of voters	⅓ of voters	⅓ of voters
1st choice	A	B	C
2nd choice	B	C	A
3rd choice	C	A	B

K Sera Sera

A random word is selected from a book in English, such as this one. Which of the following is true?

a) The chance that K is the first letter is *greater than* the chance that K is the third letter.

b) The chance that K is the first letter is *about the same* as the chance K is the third letter.

c) The chance that K is the first letter is *less than* the chance K is the third letter.

c) It is more likely that K is the third letter.

In 1973, the psychologists Daniel Kahneman and Amos Tversky wrote a seminal paper that introduced the concept of the "availability heuristic"—a way of thinking in which, when faced with a problem, we are biased toward what is more immediately familiar to us or what first comes to mind. Kahneman and Tversky's paper included an experiment in which people were shown letters and asked if, in a word randomly taken from the English language, it was more common for those letters to be in the first or the third position. They found that people showed a clear bias toward the first position, so if you answered a) in this puzzle you are in good company. We find it easier to recall words by their first letters because of a lifetime of listing words alphabetically. When thinking of words beginning with K, for example, kangaroo, kitchen, kitten, know, and, say, Kahneman may come to mind.

Yet it is hard, when put on the spot, to think of words in which K is the third letter. That would require us to stop, think again more carefully, and actually take the time to spell out words in our heads. Due to this difficulty, therefore, people tend to take a mental shortcut and jump to the conclusion that there must be many more words starting with K.

In fact, it is much more likely that K is the third letter in a randomly selected word: Words such as ask, like, likely, make, making, take, taking, bake, cake, and so on are among the most common in English.

Möbius Mind-Bender

The Möbius strip is a curious geometrical object named after the German mathematician August Ferdinand Möbius (1790–1868). It is a loop of flat material with a twist in it, which gives it the interesting property that it has only a single side. The "front" and the "back" are the same surface. So, if an ant, say, was walking along the strip, in a direction parallel to the edge, it would walk along the outside, which becomes the inside, which becomes the outside, and so on.

Here's a Möbius strip with a line drawn down the middle of its surface.

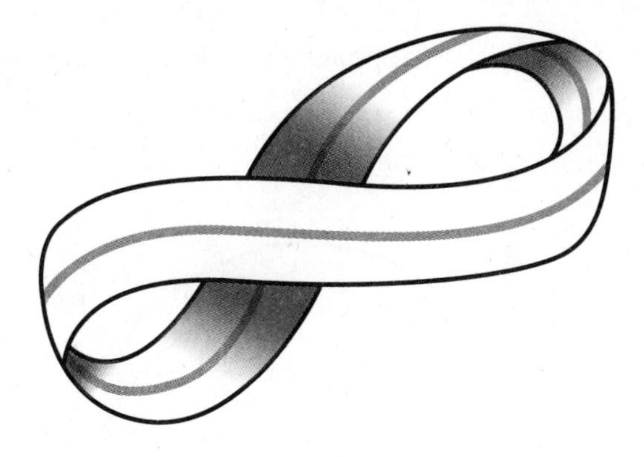

Imagine you have a Möbius strip like the one above, made of paper. What do you get when you take a pair of scissors and cut the strip in half along the line?

If you have never seen it before, it is extremely difficult to visualize what happens when you cut a Möbius strip. You might expect to see two loops, since, when cutting down the middle of a loop of flat material without a twist, that's what you would get. In fact, the cut creates a strip that is twice as long with two twists.

What is a logical mind? . . . It is the antiseptic which destroys the bacilli of unreason whereby true happiness is vivified.

WILLIAM JOHN LOCKE

Sibling Stumper

How many people is three trios of triplets thrice?

a) 9
b) 27
c) 81

b) There are 27 people.

Many people on reading this puzzle assume that a "trio of triplets" is nine people. Thus, three trios of triplets thrice is $3 \times 9 \times 3 = 81$. However, this is one of those tricky linguistic teasers.

Compare the phrase "a trio of triplets" to "a pair of twins," which is evidently only two people. A trio of triplets is three people, each one of them a triplet.

A trio of triplets thrice, therefore, is:

$$3 \times 3 \times 3 = 27$$

Spinner Winner

Here are two spinners.

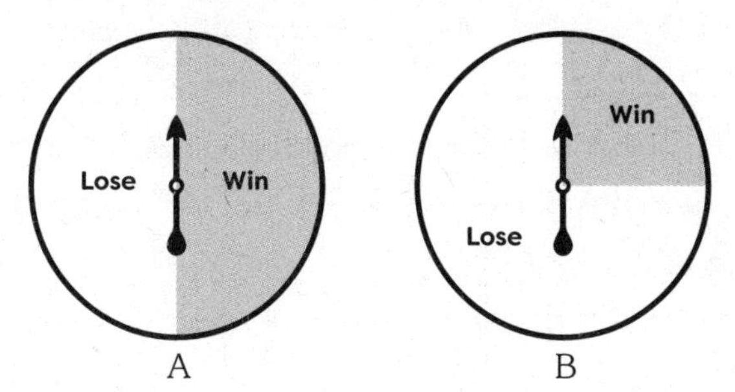

When you spin A, you have a 50 percent chance of winning, and when you spin B you have a 25 percent chance of winning.

If you land on Win after a spin, you will win $1 million.

Which of the following options is your best course of action?

a) Spin A once.
b) Spin B twice, so you get two shots.
c) It makes no difference.

a) Spin A once.

It is not always the case in life, but when gambling on these spinners, it's better to put all your eggs in the same basket. Many people think that having two shots at spinner B gives you more chances of a win. It doesn't—in fact, it gives you less chance.

If you spin A once, you have a 1/2, or 50 percent, chance of winning.

If you spin B once, you have a 1/4, or 25 percent, chance of winning, and a 3/4, or 75 percent, chance of losing. If you spin B twice, therefore, you have a $3/4 \times 3/4 = 9/16 = 56.25$ percent chance of losing both times, because the chance of two independent events occurring is the product of their probabilities. A 56.25 percent chance of losing is the same as a 43.75 percent chance of winning, so the better choice is spinning A once.

Tongue Twister

In a Canadian town, everyone speaks either English or French, or they speak both languages.

If exactly 70 percent speak English and exactly 60 percent speak French, what percentage speak both languages?

a) 30
b) 40
c) 60

a) 30 percent speak both languages.

Des MacHale, emeritus professor of math at University College Cork, says very few people get this question right because solving it requires the use of a Venn diagram, and this method of logical deduction is not often taught in schools.

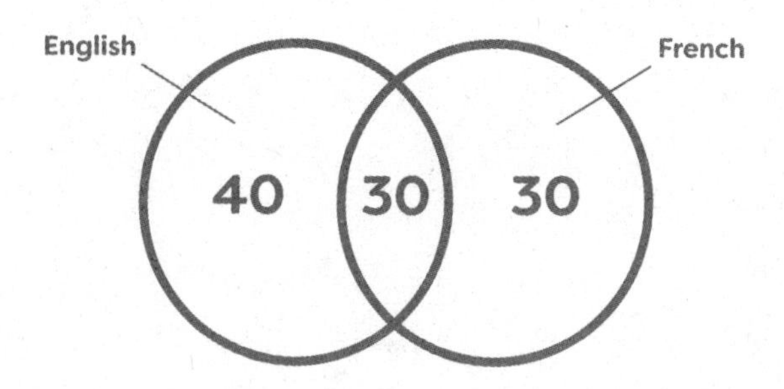

Once you draw the Venn diagram, it fills in easily. If 70 percent speak English, then 30 percent speak only French. And if 60 percent speak French, then 40 percent speak only English. Once 30 and 40 are placed in the correct regions of the diagram, the intersection containing people who speak both languages must be 30 percent.

Bamboozling Boat

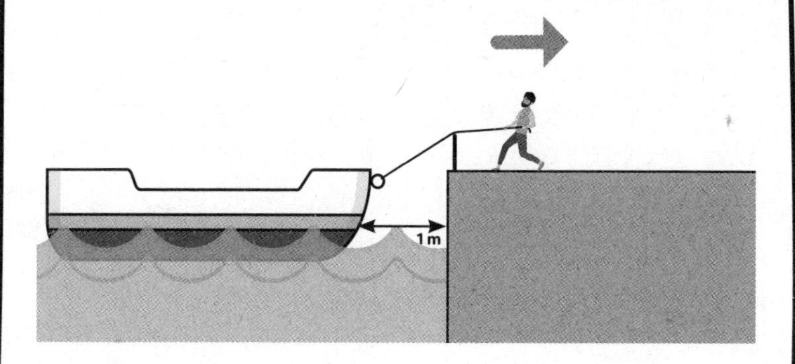

A boat is tethered to a quay by a rope, as shown above. When the rope is taut, the boat is 1 m from the quay wall.

A quayside worker wants to pull the boat closer to the wall, so they pull the rope horizontally by 1 m in the direction of the grey arrow.

Does the boat reach the wall?

a) Yes
b) No

b) Yes, the boat hits the wall of the quay, and then some!

This is very surprising. The length of rope between the quay and the boat is longer than 1 m. Reducing it by 1 m would still leave some rope left over, which gives us the erroneous impression that the boat would not quite make it to the quay. To understand the mathematics of this conundrum, we need to draw a picture.

Look at the diagram below.

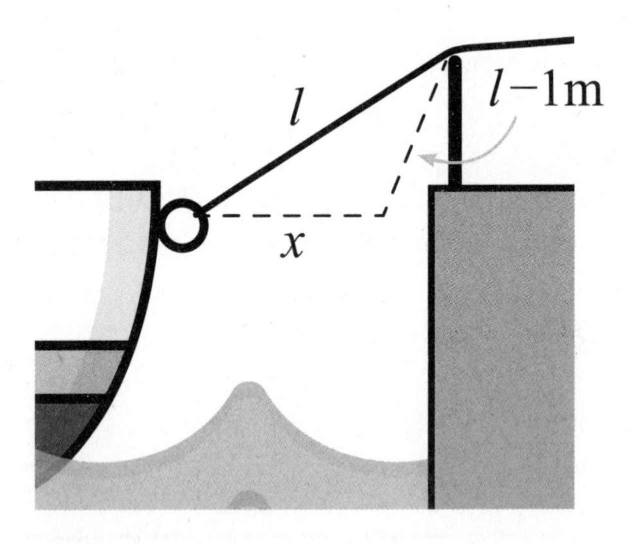

The length l is the length of rope between the post and the boat. As the quay worker pulls the rope, the boat moves toward the wall, and eventually, once 1 meter's length has been pulled, the length of rope between post and boat will be l – 1m. The horizontal distance moved by the boat is x.

In a triangle, any two sides added together are always longer than the third side. (Imagine trying to build a triangle from three sticks if this were not true.) Thus:

$$l - 1\text{m} + x > l$$

The two l's cancel each other out, leaving $-1\text{m} + x > 0$, or $x > 1\text{m}$.

But if x is greater than 1 m, which was the original distance between the boat and the wall, then the boat will smash against the quay and start lifting out of the water.

Four Kids

In a family with four children, which of the following combinations is most likely?

(Assume there is a 50/50 chance of a child being a boy or a girl.)

The children are:

a) Two boys and two girls.
b) Three of one and one of the other.

The intuitive answer is a) Two boys and two girls.

If each child is just as likely to be a boy or a girl, then, on average, after four children, you might have thought it more likely that half are boys and half are girls.

The correct answer, however, is b) Three of one and one of the other.

We can see this by listing the 16 equally probable combinations of four siblings in order of birth:

BBBB, BBBG, BBGB, BGBB, GBBB, BBGG, BGGB, GGBB, GBBG, BGBG, GBGB, GGGG, GGGB, GGBG, GBGG, BGGG

Two boys and two girls: 6/16, or 37.5 percent.

Three of one and one of the other: 8/16, or 50 percent.

Compass Confusion

In which state is the most easterly point in the United States of America?

a) Alaska
b) Florida
c) Hawaii
d) Maine

a) Alaska.

Of course, relative to the rest of the country, the most easterly point in the United States would be in Maine on the east coast of mainland America. But the question didn't ask that.

The most easterly point in the US is the eastern tip of the Alaskan island of Semisopochnoi, also known as Unyak, one of the Aleutian Islands in the Bering Sea, which is a few miles into the eastern hemisphere beyond the 180th meridian. Its coordinates place it at 179° 36' east. It is about 2,000 km west-southwest of Anchorage, but only about 1,000 km from Russia.

Baffling Bloodline

What proportion of people alive in the UK today, with predominantly British heritage, are descendants of King Edward III (1312–1377)?

Choose the option closest to your answer:

a) 5 percent
b) 25 percent
c) 50 percent
d) 100 percent

d) 100 percent.

Yes, bafflingly indeed, almost everyone with predominantly British heritage is descended from Edward III. The writer, broadcaster, and geneticist Dr. Adam Rutherford of University College London ran the sums after the cockney actor Danny Dyer appeared in 2016 on the BBC genealogy program *Who Do You Think You Are?*. When it was revealed that Dyer was a direct descendant of Edward III, the actor became emotional, saying: "I can't believe it . . . I need to get it in my nut, then move on with my life. I think I'm going to treat myself to a ruff."

It was a brilliant moment of TV, and prompted Rutherford to wonder how many other Britons of Dyer's age were also direct descendants of the same monarch.

Enlisting the help of two University College colleagues, geneticist Adrian Timpson and mathematician Hannah Fry, Rutherford followed Edward's bloodline forward as far he could, to around 1600, and then worked back from the present day, asking the question: "What is the probability that a seventeenth-century ancestor of someone born in the 1970s is in the proportion of the population who were Edward III's direct descendants?"

Edward was crowned King of England in 1327, and transformed his realm into one of the greatest military powers in Europe, initiating the Hundred Years' War with France. Aged 15, he married the 13-year-old Philippa of Hainault. They had 13 children in all: Six of them went on to have children themselves, and, according to Rutherford's count, two generations later the royal brood was 321 strong.

Estimating that each of these great-grandkids had on average two offspring, a conservative guess for that time period, the number of Edward's descendants living in 1600 would be about 20,000 in a total British population of around 4.2 million. In other words, in 1600 about one in 200 people (0.5 percent of the population) were descended from Edward. Then things get fuzzy, so Rutherford and co. decided to count back from the present day.

A person has two parents, four grandparents, and so on. Let's assume that people reproduce every 25 years, so there are four generations per century. Between 1600 and the mid 1970s, there would have been 4 + 4 + 4 + 3 = 15 generations. The number of ancestors a person has 15 generations back is 2^{15}, which is 32,768. In other words, someone born in, say, 1975, would have had 32,768 ancestors living in 1600.

The question we now want to answer is: "What are the chances that at least one of these 32,768 ancestors in 1600 would have been in the 0.5 percent of the population descended from Edward III?" An easier sum is: "What are the chances that *none* of these ancestors are in that 0.5 percent of the population?'

The calculation is $0.995 \times 10^{32,768} = 4.64^{-72}$, which is a tiny number. Essentially it means that there is no chance at all. If the chance is almost zero that *none* of the ancestors of someone born in the 1970s is descended from Edward, then the chance that *at least one of them* is descended from Edward is almost 100 percent.

Rutherford admitted that his estimate has a huge margin of error. His calculation assumes that all 32,768 of your ancestors were different people (i.e., there was no inbreeding between relatives), and that in 1600 both your ancestors and Edward's descendants were randomly distributed across the population. He adds, though, that: "The numbers come out so significantly, that even if these estimates are several orders of magnitude wrong, they still show the same answer." It is overwhelmingly likely that almost every Briton with an ancestor who predates twentieth-century immigration has a trace of Edward III's blue Plantagenet blood. And that of all his ancestors too—Edward II, Isabella of France, Edward I, and Eleanor of Castile, to name but a few.

The result is surprising because most of us commonfolk seem so far from royalty, but our bafflement is a result of a failure to grasp the rapidity of exponential growth. If a person has two children on average, and a generation is measured as 25 years, then after five centuries the twentieth generation of familial descendants comprises more than a million new births.

Coin-undrum

A 10 cm–high IKEA drinking glass is filled to the brim with water. How many pennies can I drop into it before water spills over the edge? The illustration below accurately depicts the relative sizes of the glass and a penny.

a) 1 to 5
b) 6 to 15
c) 15 to 30
d) More than 30

The answer is d) More than 30 pennies, and easily.

I know because I tried this at home using a 9 oz IKEA Pokal glass—surely one of the world's most popular drinking glasses.

On my first attempt, I managed 36 coins; I managed 39 on my second. The reason you can add a surprisingly large number of coins to a glass that is brimful of water without spillage is because of the surface tension. As the coins are added to the glass, the water's surface behaves as if it is a flexible skin, noticeably pushing upward above the brim. Eventually, however, the skin breaks and the water spills over.

What the Cluck?

If a hen and a half lays an egg and a half in a day and a half, how many eggs do half a dozen hens lay in half a dozen days?

a) Half a dozen
b) A dozen
c) Two dozen

This question became very popular in the middle of the twentieth century. It typifies a genre of puzzle that asks: "If *n* things produce *n* objects in *n* time units, how many objects would *m* things produce in *m* time units?"

For example:

> *If it takes five machines five minutes to make five widgets, how long would it take 100 machines to make 100 widgets?*

Most people get this one wrong, according to psychologist Shane Frederick, who included it as part of his "cognitive reflection test," a set of questions in which people jump to the wrong conclusions. (And which are discussed in the answer to "Years and Years," pages 84–5.) The incorrect, gut answer is 100 minutes. The correct answer is five minutes, since it takes one machine five minutes to make one widget.

OK. Back to our fractional chickens.

Assuming that all hens behave similarly, and that egg laying proceeds in a linear way, we can state that 1.5 hens lay 1.5 eggs in 1.5 days.

Four times the number of hens will lay four times the number of eggs in the same time interval—i.e., six hens lay six eggs in 1.5 days.

In an interval that is four times longer, the same number of hens will produce four times the number of eggs—so six hens will lay 24 eggs in six days.

The answer, therefore, is c) Two dozen eggs.

Logic takes care of itself; all we have to do is to look and see how it does it.

LUDWIG WITTGENSTEIN

Loopy Lineup

A crime has been committed. Here are the four suspects. As far as you are aware, each is equally likely to have committed the crime. You hire two private detectives who work independently. The first says that there is a 2–1 chance the murderer wears glasses. The second says the murderer definitely wears a hat.

Who is the prime suspect?

a) Ben
b) Bob
c) Bill
d) Bev

d) Bev!

No, the prime suspect is *not* Bob, although at first it seems most likely, since he has both glasses and a hat.

If there is a 2–1 chance the murderer wears glasses, then it is 66.66 percent likely that the murderer is Ben, Bob, or Bill. In other words, the likelihood of each of these suspects being the murderer is 22.22 percent, and the chance that Bev is the murderer is 33.33 percent. If the murderer definitely wears a hat, the prime suspect must be either Bob or Bev. We've seen that Bev is more likely than Bob, so Bev is the prime suspect.

Snow Problem

Which is the more likely event?

a) The school is closed today. Children must stay at home.

b) The school is closed today because a snowstorm in the night left all roads perilous for pedestrians. Children must stay at home.

The correct solution is a) The school is closed today. Children must stay at home.

We can see the problem in terms of a Venn diagram. The set of school closures due to snowstorms is contained within the set of all school closures, which could be due to any number of reasons: snowstorms, gas leaks, a plague of rodents, whatever.

Thus, b) is less likely than a).

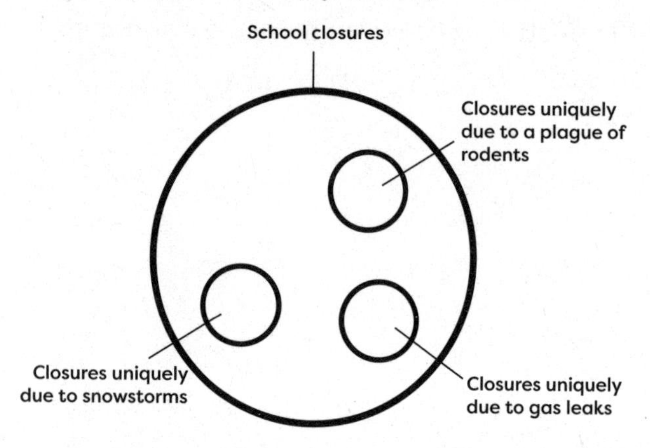

Yet, while a) is clearly more likely than b), people tend to say that b) is more likely because it paints a plausible picture—we can visualize the snow-covered footpaths and the perilous patches of black ice sending the school's shivering pupils crashing to the ground. Our minds overrule the mathematics to go with what feels like a good story.

The phenomenon whereby people tend to judge the probability of two conjoined events as being more likely than the probability of one of them happening, even though this violates the laws of probability, is known as the "conjunction fallacy." It is also known as the "Linda problem" because of the most famous example of the genre, devised by psychologists Daniel Kahneman and Amos Tversky in 1981:

> *Linda is 31 years old, single, outspoken, and very bright. She majored in philosophy. As a student, she was deeply concerned with issues of discrimination and social justice, and also participated in anti-nuclear demonstrations.*
>
> *Which scenario is more likely to be true:*
>
> *a) Linda is a bank teller.*
>
> *b) Linda is a bank teller and is active in the feminist movement.*

Kahneman and Tversky said that most people responded b).

Now, which of the following is more likely?

a) You are reading this book.
b) You are reading this book, and I'm hoping you'll get the next puzzle wrong.

Onward!

Tricky Test

Lurgy-X is a horrible disease that currently affects one person in 10,000.

There is a very accurate test for Lurgy-X that always shows positive if you have the disease.

However, very occasionally, in a tiny number of cases, the test is unreliable. One percent of tests produce a false positive: That is, the test declares you have the disease when you don't.

You test positive for Lurgy-X.

Which is true?

a) It is extremely likely you have Lurgy-X.
b) It is quite likely you have Lurgy-X.
c) You are as likely to have Lurgy-X as not to have it.
d) It is quite unlikely you have Lurgy-X.
e) It is extremely unlikely you have Lurgy-X.

e) It is extremely unlikely you have Lurgy-X.

If you tested positive for a horrible disease like Lurgy-X, you would naturally feel very worried. However, if the false-positive rate is 1 percent, the chances you have it are very small indeed.

A false-positive rate of 1 percent means that in 10,000 tests of people without the disease, 100 people will be told they have it. I stated in the question that one person in 10,000 has Lurgy-X, so in any group of 10,000 you can expect one person to have it. So the chances that you have the disease are about one in 100, or 1 percent.

Or to put it the other way round, you are 99 percent likely to be free of the disease.

Silly Sentence

How many mistakes are there in the following sentence?

This sentence is kurious because it it only contayns three misteaks.

There are five mistakes in total.

Did you get them all? The puzzle was designed to conceal different types of mistake. When you are looking for one type of mistake, other things can easily slip through.

There are three obvious spelling mistakes (kurious, contayns, and misteaks). There is the repetition of "it," and an incorrect statement: There are *not* three mistakes, there are five.

Ropey Problem

A 101-meter piece of rope is tethered to the ground at two posts 100 meters apart. Midway between the posts, suppose we take the midpoint of the rope and hoist it directly upward until it becomes taut, leaving a gap underneath for a vehicle to drive through.

Of the vehicles listed below, which is the largest that would be able to drive through without touching any rope?

a) A go-kart
b) A car
c) A bus
d) A double-decker bus
e) A double-decker bus with an SUV on top

The answer is e) A double-decker bus with an SUV on top, which surprises many people.

Our intuitions about space are often wrong. Even though the rope has only 1 m slack, which is 1 percent of the distance between the posts, it can be raised to more than 7 m in the middle. A double-decker bus is about 4.4 m high, and an SUV about 1.8 m, so together at 6.2 m they will easily fit under the rope.

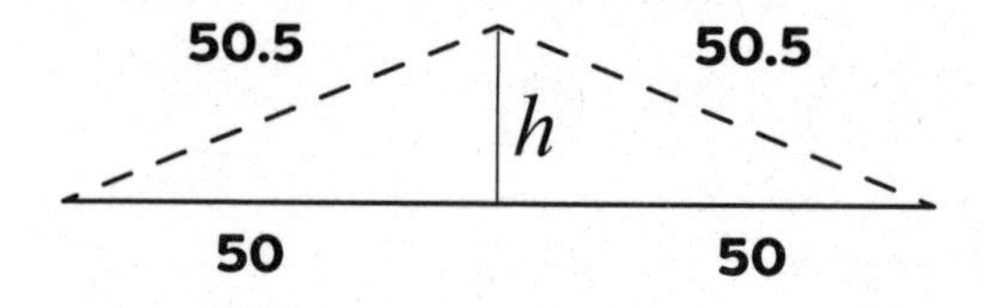

We solve this problem using the Pythagorean theorem, which states that for right-angled triangles, the square of the hypotenuse is equal to the sum of the squares of the other two sides. The diagram above shows the rope lifted in the middle: It makes two right-angled triangles. Each has a hypotenuse of half the rope length, 50.5 m; one side of half the distance between the posts, 50 m; and the other side the height *h*.

Thus: $h^2 + 50^2 = 50.5^2$

$h^2 + 2500 = 2550.25$

$h^2 = 2550.25 - 2500 = 50.25$

$h = \sqrt{50.25} = 7.1$

"Contrariwise," continued Tweedledee, "if it was so, it might be; and if it were so, it would be; but as it isn't, it ain't. That's logic."

LEWIS CARROLL

Years and Years

Aaron and Betsy have a combined age of 50.

Aaron is 40 years older than Betsy.

How old is Betsy?

Betsy is five years old.

If you guessed that Betsy is 10 years old, you are not alone. Most people do.

The question is a version of what is perhaps the most famous "puzzle that almost everyone gets wrong." In its original formulation, in a 2005 paper by psychologist Shane Frederick, we are asked to compare the price of sporting paraphernalia:

A bat and a ball cost $1.10.
The bat costs $1.00 more than the ball.
How much does the ball cost?

Frederick found that most people respond swiftly, but incorrectly, with $0.10. They are misdirected by the $1.10 and $1.00 and subtract one from the other. In fact, the ball costs $0.05, and the bat costs $1.05.

Likewise, in the case of Aaron and Betsy, one's gut reaction may be that Betsy is 10, when in fact she is five, and Aaron is 45.

Frederick called his problem a "cognitive reflection test" because the point is not to assess mathematical ability (the arithmetic is easy), but to investigate our ability to stop and think. If you answer impulsively, you are likely to fall into the trap and go for the tempting answer of $0.10. But if you stop for a moment and reflect, you will see that the answer is $0.05.

(In his 2005 paper, Frederick includes two other cognitive reflection tests, and I have posed versions of both of them at other places in this book.)

Daniel Kahneman popularized the bat-and-ball problem in his book *Thinking, Fast and Slow*, to illustrate his theory that the mind has two systems: one that works quickly and intuitively but is easily misled, and one that is slower, reflective, and more analytical. The first system reaches the easy, incorrect answer, but the second will eventually get it right. (The reason I gave a question about ages, rather than prices, is that I assumed that many readers would already be familiar with the bat-and-ball problem.)

There is a vast body of literature on Frederick's cognitive reflection tests. In 2019, Gordon Pennycook, a psychologist at Cornell University, demonstrated that people who get the bat-and-ball problem wrong are also more susceptible to partisan fake news. His conclusion was that believing implausible headlines is more a result of lazy thinking than partisan bias.

In 2023, Andrew Meyer and Shane Frederick looked at many subtle variations of the bat-and-ball problem—and some not-so subtle ones—in which the correct answer is suggested. Such as:

> *A bat and a ball cost $110 in total. The bat costs $100 more than the ball.*
>
> *How much does the ball cost? Before responding, consider whether the answer could be $5.*

More than half of respondents—55 percent—still gave the answer $10.

This next version is more prescriptive:

> *A bat and a ball cost $110 in total. The bat costs $100 more than the ball.*
>
> *How much does the ball cost? The answer is $5. Please enter the number 5 in the blank below.*
>
> $ ____________

Even after being told the answer and asked to write it down, 18 percent of people still responded with $10.

According to Meyer and Frederick, we should perhaps categorize people into three types: the *reflective* (who get the bat-and-ball question correct), the *careless* (who get the wrong answer at first but will update their answer when they are told it is wrong), and the *hopeless* ("who are unable or unwilling to compute the correct response, even when told that $10 is *not* the answer").

When everything is easy, one quickly gets stupid.

MAXIM GORKY

Spot the Difference

What is the most striking difference between these two pictures—other than the obvious fact that one is upside down?

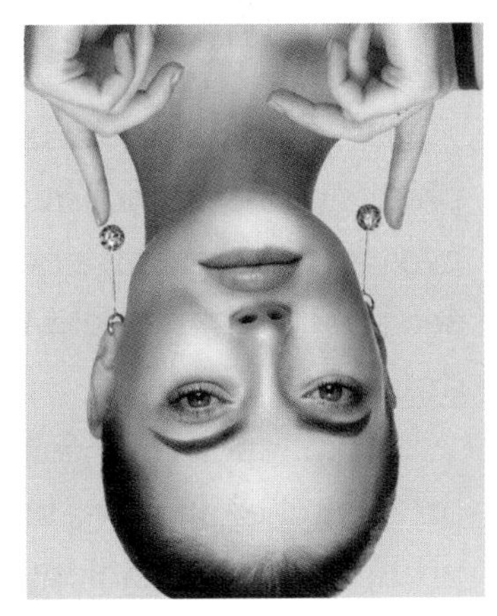

When looking at the book the right way up, the most obvious difference seems to be the color of the earrings. Superficially, not much else stands out as unusual.

Now turn this book upside down.

Eeeek!

The striking anomalies are clearly the woman's eyes and mouth, which have been turned upside down on her face. She has been transformed into a gargoyle!

When looking at an upside-down portrait or photograph, it is not instantly obvious that the eyes and mouth have been inverted, but when you see the image the right way up, the effect is quite dramatic.

This phenomenon is called the "Thatcher effect" or "Thatcher illusion," since Peter Thompson, professor of psychology at the University of York, who first demonstrated it in 1980, originally used an image of Margaret Thatcher.

Bemusing Balls

How many different colors of ball are there?

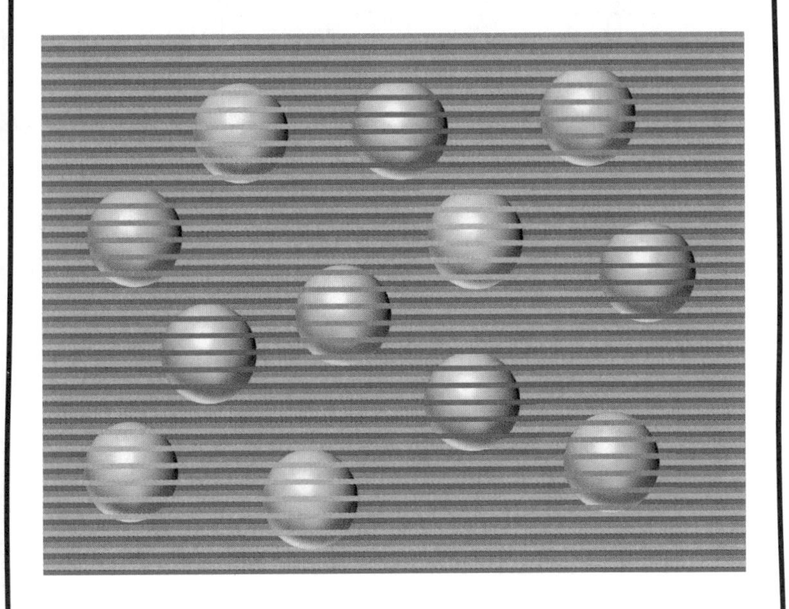

You may have thought you saw balls of three different colors—lime green, orange, and lilac—but in fact they are all the same beige color.

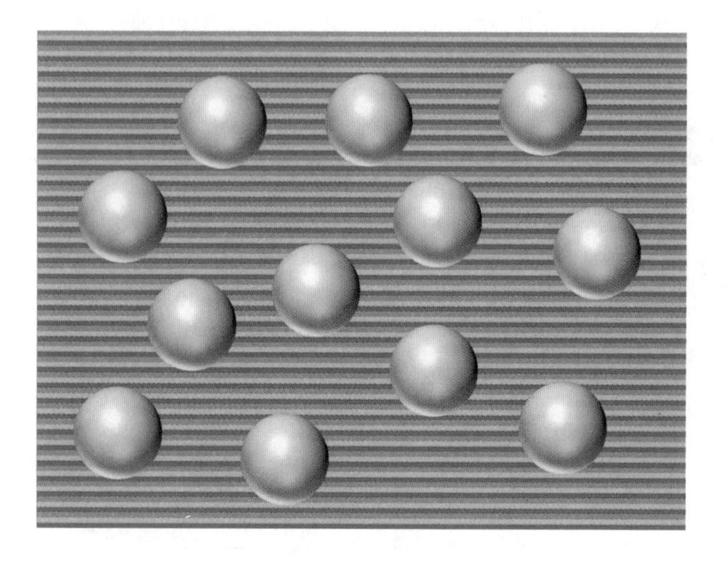

This image, known as the "confetti illusion," is probably the most extraordinary version of many similar optical illusions in which the neighborhood adjacent to a shape influences our perception of that shape's color. In this case, the clever combination of colored stripes gives the illusion of three different colors of ball.

Card Sharp

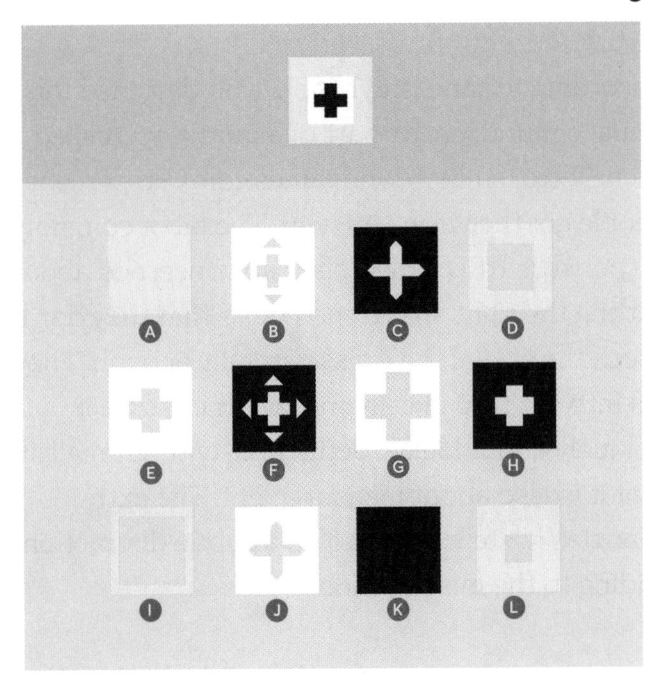

The shapes A to L above are either stencils or colored squares. Imagine they are made out of card.

The top image is created by placing some of the shapes on top of each other.

Which ones, and in which order?

You need to place stencils KBD on top of each other, in that order.

According to Jane Braybrook, who designed this visual conundrum as part of a puzzle app called Stenciletto (smileyworldgames.com), 90 percent of people get the wrong answer. The most common response is KED. Solvers are so concerned about getting the right shape and colors that they don't check the size of the cross carefully enough. They fall into the trap of thinking that the puzzle is uniquely about logical deduction, without realizing that it is also about measurement. The extra triangles in stencil B are also a subtle distraction, adding to the misdirection.

Stenciletto is based on an IQ test devised in the 1920s by the American psychologist Grace Arthur.

Pavement Poser

What black material is on the pavement, outlined by chalk?

a) Oil
b) Chalk
c) Felt
d) Something else

The answer is d) Something else—or rather, nothing at all!

The black "material" is simply the shadow of the posts. But when shadows are outlined in chalk, they can appear so deeply black that they look like a physical material.

The American artist, printmaker, and photographer Michael Neff captured this image, which comes from "The Night Shadow," an ongoing project in which he outlines shadows of city-street objects at night with chalk and takes pictures of the results. (See michaelneff.com.)

Broken Line

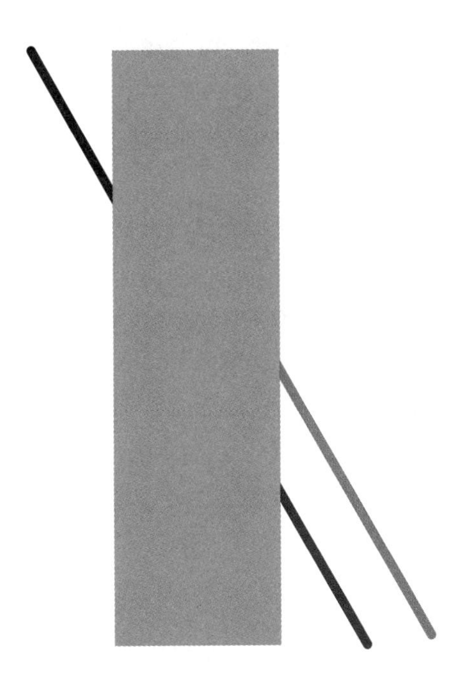

Does the black line at the top left line up with the red line or the blue one?

It lines up with the lower, red line.

On first look, however, one's hunch is that the higher, blue line is the correct answer, although this misperception can be quickly checked with a ruler.

In 1860, Johann Christian Poggendorff, the German physicist and longtime editor of Europe's foremost scientific journal, *Annalen der Physik und Chemie*, noticed that when a diagonal line is interrupted by a solid vertical column, the brain seems to shift the position of the line segments vertically on either side so they don't appear to be aligned, even though they are.

This optical trick is now known as the "Poggendorff illusion."

Murky Med

Here is a map of the Mediterranean.

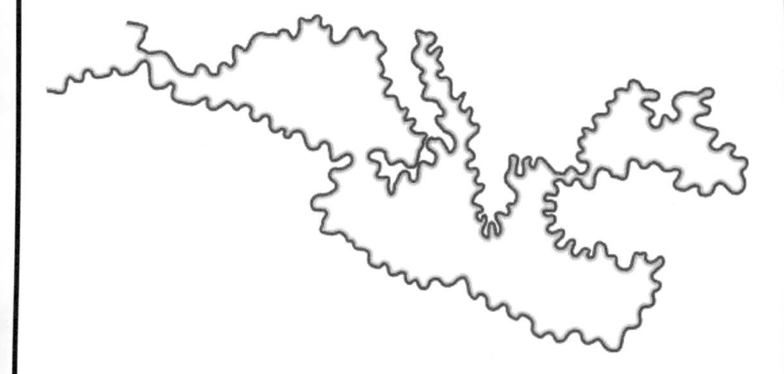

What color is the sea?

The Med is white.

The outline of the coast is purple and orange, but apart from that there is no color in the image, even though it clearly looks as if the sea is painted in a washy orange tint.

This odd effect—two colors along a border making the white area next to it appear shaded—is known as the "watercolor illusion." The phenomenon was discovered in 1987 by the Italian psychologist Baingio Pinna, a professor at the University of Sassari, and one of the foremost specialists working in the field of visual perception today. As well as giving the white area color, the illusion also creates a kind of three-dimensional effect: In the image, the Mediterranean Sea could almost be a dribble of oil spread across the page.

Café Confusion

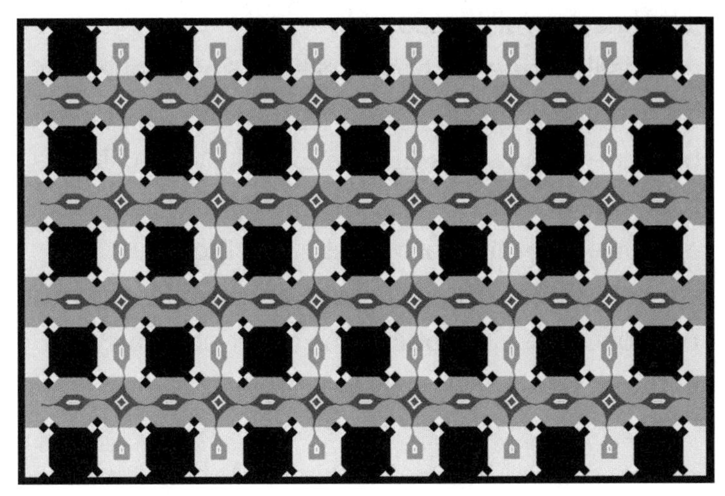

At what angle to the horizontal are the dark blue strips?

a) 0 degrees
b) 5 degrees
c) 45 degrees

a) 0 degrees.

The lines are all horizontal.

This illusion—that parallel lines don't *appear* to be parallel when separated by alternately colored bricks—is known as the "Café Wall illusion." It was first described in 1973 by the British psychologist and professor of neuropsychology Richard Gregory (1923–2010), after one of his colleagues at Bristol University noticed the effect on a café wall in the city.

Fool House

You are playing poker with friends. Which of the following two hands is stronger?

For those unfamiliar with the rules of poker, an ace is the highest-ranked card, followed by king, queen, jack, 10, 9, and so on down to 2.

The best hand is a ***royal flush****, which is ace, king, queen, jack, and ten in the same suit—all spades or all hearts, for example.*

The second-best hand is a ***straight flush****, which is any run of five cards of successive values in the same suit that is not a royal flush.*

The third-best hand is ***four of a kind****, which includes the four cards of the same value in each of the four suits—spades, clubs, hearts, and diamonds.*

And the fourth-best hand is a ***full house****, which is three of a kind (three aces, three jacks, or three threes, for example) and two of another.*

Both hands in this puzzle are full houses.

Hand b) is stronger, surprisingly, even though its individual cards are weaker.

This is because only one pack of 52 cards is used in any game of poker. Therefore, if you have hand a), no one else can have hand b), and vice versa, so they are not competing against each other. Both hands will beat all possible fours of a kind, because no other hand can have three aces. Both hands will lose if the opponent has the only possible royal flush (in clubs). But hand b) eliminates more chances that your opponent has a straight flush. For example, if you have hand b), you have the 9 of hearts and of spades, which means your opponent cannot have a run of cards in hearts or spades of 5 to 9, 6 to 10, 7 to jack, or 8 to queen. If you have hand a), the only straight flush in hearts or spades that your opponent *cannot* have is 9 to king.

Since hand a) gives your opponent more ways to beat you, it is the weaker hand.

Slinky Thinkie

One end of a Slinky helical spring is held in the air. A weight is attached to the other end, which pulls the Slinky vertically down.

When the person holding the Slinky lets go of it, what first happens to the weight?

a) It goes up.
b) It does not move.
c) It falls down.

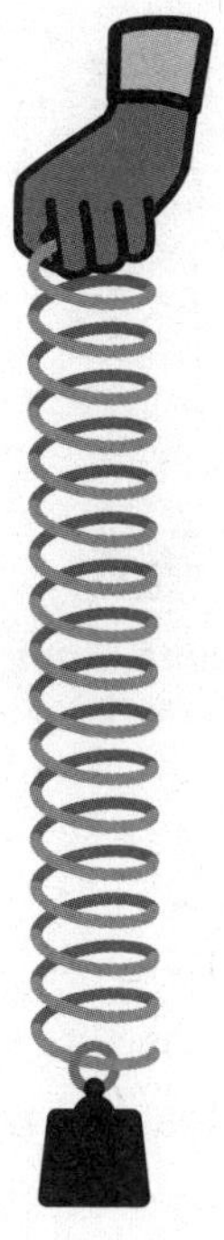

b) The weight does not move. It briefly stays suspended in the air.

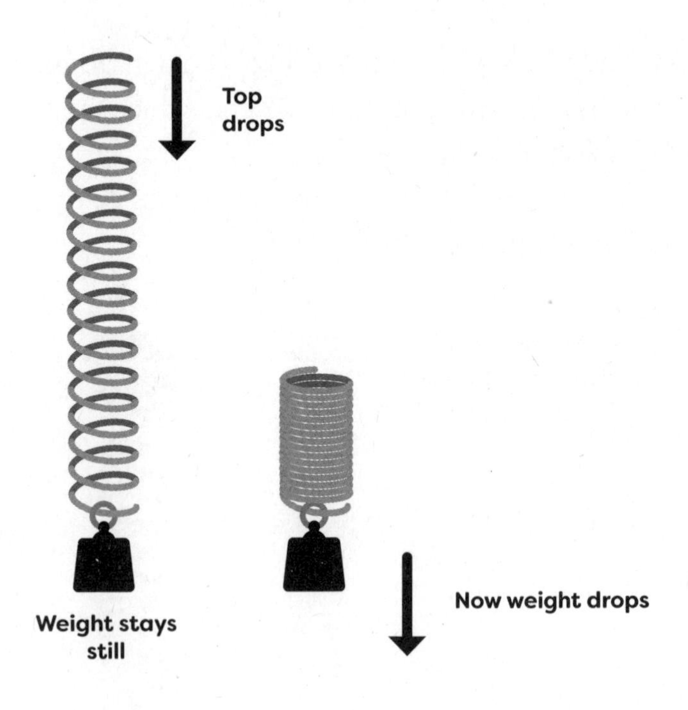

When the person holding the Slinky releases it, the upper, previously taut coils start to fall, releasing the tension in the spring as it moves down. Only once the Slinky has lost all its tension and is "at rest" does the weight (together with the Slinky) begin to fall to the ground.

One way to think about this situation is as follows: At the beginning, the weight is being held up by the tensile force in the bottom of the helical spring. When the top end of the Slinky is released, this force is still present in the bottom end of the Slinky. And it takes a brief moment for the message that the Slinky is dropping to travel down through the spring to the weight.

The same phenomenon occurs even when there is no weight attached to the bottom end of the Slinky. When you hold a Slinky from the top and then drop it, the bottom end briefly stays suspended in mid-air.

If you don't believe me, try it out at home, or look online: There are many excellent clips on YouTube of surprising Slinky science.

Money Month

Which of the following will make you richer?

a) Over the next 30 days, I'll give you $10,000 a day.

b) Over the next 30 days, I'll give you 1¢ on the first day, 2¢ on the second day, 4¢ on the third day, 8¢ on the fourth, and so on, doubling each time.

The most lucrative option by far is b) Over the next 30 days, I'll give you 1¢ on the first day, 2¢ on the second day, 4¢ on the third day, 8¢ on the fourth, and so on, doubling each time.

With option a), after 30 days you will have amassed $300,000.

If you go for option b), on the other hand, on day 30 you will receive $5.4 million, giving you a total for the month of more than $10 million. (On day *n* you will earn 2^{n-1} cents, and 2^{29} = 536,870,912.)

The daily values in b) may start out very small, but they grow exponentially, so the amount of money amassed quickly accumulates. Most people tend to underestimate exponential growth and, without taking the time to work out the sums, may judge that 1¢ doubled 30 times is likely to be less than $10,000 multiplied by 30. In fact, it is much more.

Dirty Dozen

What is 12 divided by a half?

The answer is 24.

Many people get confused and think that dividing by a half is the same as dividing by 2, when, in fact, it is the same as multiplying by 2.

Jolly Good Fellows

One of the most famous results in probability is the amazing fact that in any random group of 23 people, the likelihood that two people share the same birthday is just over 50 percent. So, for example, whenever you are watching a soccer match, it is more likely than not that two people on the field (of the 22 players and the referee) share the same birthday. The result is called the "birthday paradox" because it seems to go against common sense: 23 is a remarkably small number of people for a shared birthday when there are 365 potential birth dates to choose from.

Now to the question: To be more than 99.9 percent sure that two people in a group share the same birthday, what's the smallest number of people we need?

a) 70 people
b) 100 people
c) 150 people
d) 200 people

The answer is a) 70 people.

Again, a counterintuitively small amount. I find it remarkable that even though there are 365 days in a year, you only need a group of about 70 people to be almost certain that two of them share a birthday. The moral is: Coincidences are not only likely in life, they are pretty much guaranteed!

To work out the chance of a shared birthday in a group of random people, you first need to work out the chance of *no* shared birthdays. Let's start with a group of two people.

The chance of no shared birthdays in a group of two is 364/365, or 99.7 percent, because once you have established the birthday of the first person in the group, there are 364/365 chances that the second person was born on a different day.

With a group of three people, the chance of no shared birthdays is 364/365 × 363/365, or 99.2 percent. With a group of four it is 364/365 × 363/365 × 362/365, or 98.4 percent, and so on. For each new person in the group, the pool of available dates for different birthdays reduces by one.

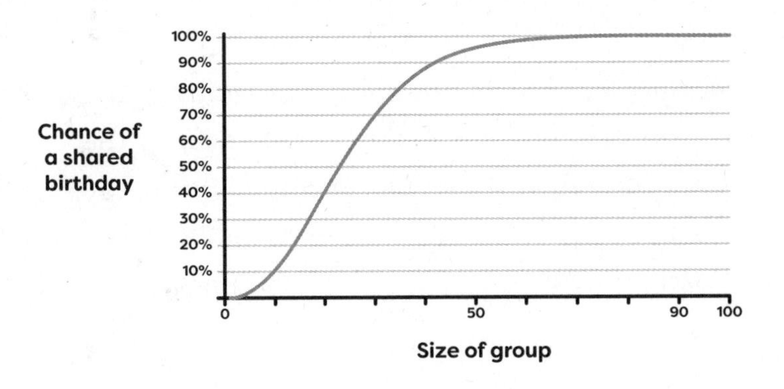

Looking at the situation in terms of shared birthdays, in a group of two the chance of a shared birthday is 0.3 percent; in a group of three it is 0.8 percent; and in a group of four it is 1.6 percent. Slowly the percentage is increasing. I'll spare you the calculations, but by the time the group has 23 people in it, the chance of a shared birthday exceeds 50 percent for the first time—50.7 percent—which means that the chance of a shared birthday within a 23-person group is more likely than not.

If you carry on increasing the size of the group, the chance of a shared birthday rockets very quickly, as you can see in the graph above. At 30 people, you reach a

70 percent chance of a shared birthday, and at 41 people there is a 90 percent chance. By the time you reach 70 people, as stipulated in the question, the chance of a shared birthday is 99.91 percent.

Head-Spinner

Suppose you place a quarter in the bottom position below and keep it in place with your finger. Put a second quarter in the top position and roll it around the first coin counterclockwise.

When the rolling coin returns to its original position, how many times has Washington's head rotated?

Washington's head rotates twice.

Try it! If you have never done it before, it is very surprising that the head does a double rotation, when you might assume it rotates only once. This is because the coin is actually doing two rotations: one around itself, and one around the other, static, coin.

Shoot-Out Shenanigans

At the end of the cup final, United beat Rovers 5–3 on penalties.

Who took the first penalty?

a) United
b) Rovers
c) Impossible to say

(When a soccer match in a knockout competition ends in a draw, after extra time it progresses to a penalty shootout in which the teams alternately take five penalty kicks against each other. The score mentioned above is the tally of scored penalties in the shootout.)

a) United took the first penalty.

The intuitive answer is that you don't have enough information to solve the problem, but in fact you do, because some penalty shootout scores are impossible.

If United won 5–3 against Rovers, they scored all their penalties. If United were the second team to take their first penalty shot, this would mean that the Rovers had taken all their penalties before United's final penalty—at which point the score was 4–3 United. Yet if the score was 4–3 United, and Rovers had already taken all their penalties, the game would have ended there and then, since Rovers had already lost.

Thus 5–3 is an impossible score when the winning team is the second to take the penalties.

United must have gone first.

Just the Job

Phyllis is a quiet-living Briton who likes classical music, does a Sudoku every day, and wears glasses.

Which is more likely?

a) Phyllis is a supermarket worker.
b) Phyllis is a school librarian.

a) It is much more likely that Phyllis works in a supermarket.

While it is true that the percentage of supermarket workers who like classical music, do a Sudoku every day, and wear glasses is probably lower than the comparable percentage of school librarians, there are many more supermarket workers (almost a million in the UK) than there are school librarians (maybe around 10,000). The absolute number of supermarket workers with these attributes is therefore likely to be considerably greater than the number of school librarians with them.

This puzzle is a perfect example of our propensity to be unduly influenced by personal details and jump to the more obvious conclusion, ignoring the relevant underlying statistical information—a phenomenon known as "base-rate neglect."

Curious Counters

You have three counters:

One is black on both sides.
One is white on both sides.
One is black on one side and white on the other.

The counters are placed in a hat in a random order. You close your eyes, put your hand in the hat, take out a single counter, and put it on the table.

The face-up side is black.

Which of the following is true about the face-down side of this counter?

a) It has a 50/50 chance of being black or white.
b) It is more likely to be white.
c) It is more likely to be black.

The correct answer is c) The reverse side of the counter is more likely to be black.

However, most people answer a) It has a 50/50 chance of being black or white, reasoning that if the face-up side is black it must be one of two counters: counter 1 (black/black); or counter 3 (black/white). In the former case the reverse is black, and in the latter case the reverse is white. Since it's a choice between two counters, according to this line of deduction, the chance of black or white on the opposite side must be 50/50.

Yet this reasoning is fallacious. The correct way to think about this problem, and many others in probability theory, is to consider *all* the equally possible events. If a counter is taken out of the hat to reveal a black side face up, there are *three* equally likely possibilities, because three sides of the counters are black.

The black face-up side, therefore, could be one of these three options:

1. One side of counter 1
2. The reverse side of counter 1
3. The black side of counter 3

In two of these scenarios, the reverse side of the counter is black, and in one scenario the reverse side of the counter is white. Thus, if the face-up side is black, the chance that the other side is black is 2/3, or 66.6 percent. So, it is much more likely that the reverse side is black.

This puzzle and others like it are much studied in psychology. The researchers Maya Bar-Hillel and Ruma Falk wrote about this version of the problem in their 1982 paper *Some Teasers Concerning Conditional Probabilities*. In it they stated that about two thirds of respondents thought that the chance of the counter being black on the other side was 50 percent.

The problem itself, however, is much older, dating back to the nineteenth century. In its original form it is known as "Bertrand's box paradox," after the French mathematician Joseph Bertrand (1822–1900), who wrote about it in his 1889 book *Calcul des Probabilités*. Rather than counters, he stated the puzzle in terms of three boxes: One box contained two gold coins, one two silver coins, and one contained one of each coin. You take a box at random and then select a coin at random from that box. If the coin you select is gold, what are the chances the other coin in that box is also gold?

Bertrand's box paradox is the predecessor to many famous probability conundrums, including perhaps the most famous of them all, the "Monty Hall problem." Monty Hall was the host of the game show *Let's Make a Deal*, which featured a segment in which a prize was hidden at random behind one of three doors. Here's the puzzle:

> *You are on* Let's Make a Deal. *In front of you are three doors. Behind one door is an expensive car; behind the other two doors are goats. Your aim is to choose the door with the car behind it. The host, Monty Hall, says that once you have made your choice he will open one of the other doors to reveal a goat.*
>
> *The game begins and you pick door No. 1.*
>
> *Monty Hall opens door No. 2 to reveal a goat.*
>
> *He then offers you the option of sticking with door No. 1 or switching to door No. 3.*
>
> *Is it to your advantage to make the switch?*

The common response is that it makes no difference whether you switch or not. There are two doors left, so it's 50/50 whether the prize car is behind either. (Just as you may have reasoned that it is 50/50 with the counters in the puzzle above.)

However, the correct answer is that switching doubles your chances of winning the car!

Consider all equally possible events. At the beginning of the game, there is a one-in-three chance the car is behind door 1, a one-in-three chance the car is behind door 2, and a one-in-three chance the car is behind door 3.

Therefore there is a combined two-in-three chance that the car is *not* behind door 1. Once Monty reveals the goat behind door 2, the chance that the car is *not* behind door 1 is still two in three, so the chance that the car *is* behind door 3 rises to two in three.

Rain-teaser

Where is the world's largest desert?

No, the world's largest desert is not in Africa. The Sahara is the world's largest *hot* desert, but the question did not specify temperature.

A desert is an area that receives very little rainfall, usually categorized as less than 25 cm a year. Antarctica (14.2 million square kilometers) and the Arctic (13.9 million square kilometers) are respectively the world's largest and second-largest deserts. The Sahara (9.4 million square kilometers) is therefore only the third-largest desert in the world.

Möbius Mind-Twister

Here's a Möbius strip with parallel lines drawn down the middle of its surface, separating the strip into equal thirds.

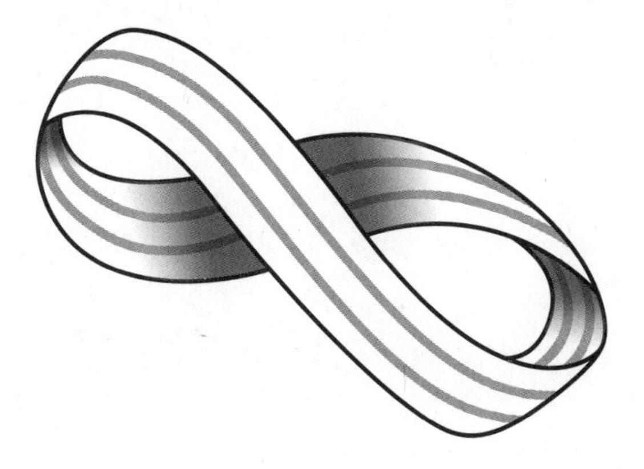

What do you get when you cut along the lines?

You get a Möbius strip which is twice as long with two twists (just as you did in "Möbius Mindbender," page 38), but this time it is interlocked with another Möbius strip.

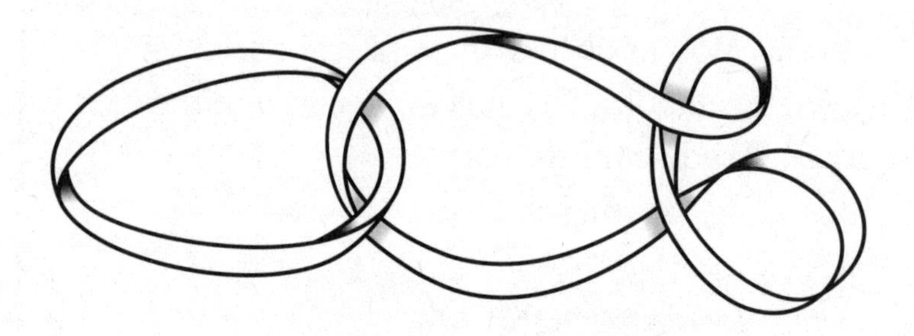

Deck Dilemma

I have a standard deck of 52 playing cards. I randomly take a card, look at it, and keep it concealed from you.

Your task is to guess my card.

Before you guess, you can ask me one of these three questions, which I will answer truthfully:

a) Is the card red?
b) Is the card a jack, queen, or king?
c) Is the card the ace of spades?

Which question should you choose in order to have the best chance of guessing my card?

It doesn't matter.

Why? Well, before you ask any of the three questions, you have a 1 in 52 chance of naming the correct card. And *after* you ask any of the questions, you have a 1 in 26 chance of naming the correct card. Although it seems counterintuitive, whatever question you ask improves your chance of guessing correctly by exactly the same amount.

To see how the math works, let's take it question by question.

Question a) Is the card red? Once I reply, you will know if my card is red or black. There are 26 red and 26 black cards, so you have a one in 26 chance of guessing the correct one.

Question b) Is the card a jack, queen, or king? There is a 12/52 chance the card is a jack, queen, or king, and a 40/52 chance it isn't. If I tell you it is a face card, you have a 1 in 12 chance of guessing correctly, and if I tell you it isn't, you have a 1 in 40 chance.

The chance of guessing the correct card is the chance of guessing it when it is a jack/queen/king

(1/12 × 12/52 = 1/52), plus the chance of guessing it when it is *not* a jack/queen/king (1/40 × 40/52 = 1/52). The sum of these two probabilities, 1/52 + 1/52, is 1 in 26.

Question c) Is the card the ace of spades? The same argument applies. There is a 1 in 52 chance of the card being the ace of spades, and a 51/52 chance of it being any other card. If the correct card is not the ace of spaces, you have a 1 in 51 chance of guessing it. Thus, the chance of guessing correctly when the card is *not* the ace of spades is 51/52 × 1/51 = 1/52. Again, the sum of both possible outcomes is 1/52 + 1/52 = 1/26.

Doubt is
the origin
of wisdom.

RENE
DESCARTES

Pair-plexing Problem

Your sock drawer contains 38 individual socks, of which 10 are red, 11 are green, and 17 are white. The light in your room is broken and you cannot identify the socks' colors when they are in the drawer.

How many individual socks do you need to take out of the drawer to be sure you will have a pair of the same color?

You only need to take out four socks.

If the first three are all different colors, the fourth will give you a pair. If the first three are not all different colors, you have a pair already!

Often people misunderstand what is being asked for and give a much larger number.

Pointed Question

A geometrical "point" is an exact location in space. It has no length, width, or height. Are there the same number of geometrical points in a line 3 cm long and a line 6 cm long?

3 cm line ____________

6 cm line ________________________

a) Yes
b) No

The intuitive answer is that there are twice as many geometrical points in a line twice as long.

Certainly, there is twice as much ink in a line twice as long, but since this book revels in counterintuitive answers you won't be surprised to discover that the answer is a) Yes—both lines have the same number of points. The fact that lines of different length have the same number of geometrical points is more than just a puzzle: It is an introduction to the wonderful world of mathematical infinity.

First, let's answer the question.

Consider the diagram on the opposite page, in which the 3 cm line is above the 6 cm line. I have added a point A above both lines, such that a line drawn from point A that touches the left end of the 3 cm line also touches the left end of the 6 cm line, and that a line drawn from A that touches the right end of the 3 cm line also touches the right end of the 6 cm line. (These lines are dashed.)

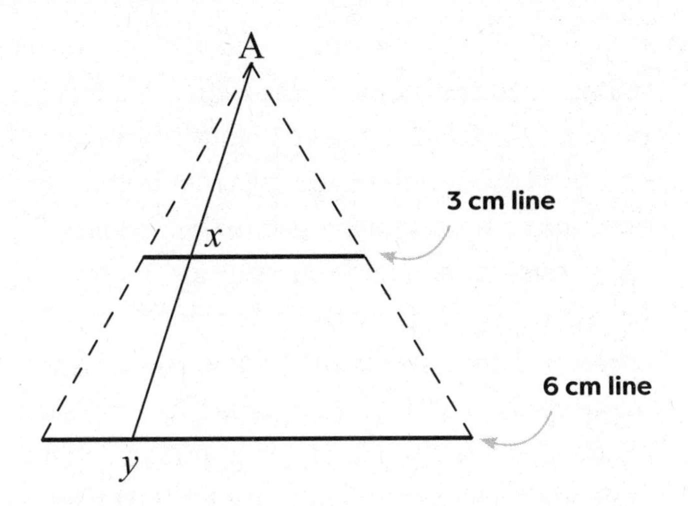

To show that the 3 cm line has the same number of points as the 6 cm line, we need to show that for every point on the 3 cm line there is a unique point on the 6 cm line, and vice versa. So, let's take a point at random on the 3 cm line, and call it x.

By drawing a line from A through x we can determine a unique point on the 6 cm line, y. Likewise, from any point y on the 6 cm line, we can draw a line to A that determines a unique point on the 3 cm line. Consider the line from A as a movable needle—at any position we move it to, it will contain two unique points on both lines.

If every geometrical point on the 3 cm line has a

unique "partner" point on the 6 cm line, and every point on the 6 cm line has a unique "partner" point on the 3 cm line, we can deduce that both lines have the same number of geometrical points.

Another question now arises: How many geometrical points *are* there on a line that is 3 cm long? The answer (which came about after major conceptual breakthroughs in the nineteenth century) is that there is an "uncountable infinity" of geometrical points. The name for the infinite number of points on a line is c, short for the "continuum." Every line, whether it's 3 cm, 6 cm, or indeed 6,000 miles long, has c points. The number c is provably larger than the infinity we learn at school, commonly referred to as ∞, which is a "countable infinity" since it is what we get when we start counting 1, 2, 3 and never stop. The difference between ideas like ∞ and c, and indeed a whole hierarchy of differently sized infinities, is one of the core research areas in the philosophy of mathematics.

Turnover Task

Each of the following four cards has a letter printed on one side and a digit on the other.

E	Q	3	6

Which card(s) must you turn over in order to check that every card with a vowel on one side has an odd digit on the other side?

You need to turn the E card and the 6 card.

The trap most people fall into is thinking that you need to turn over the 3 card. However, you don't.

If the 3 card does have a vowel on the other side, then it satisfies the rule "every card with a vowel on one side has an odd digit on the other side."

If the 3 card does *not* have a vowel on the other side, the card does *not* break the rule "every card with a vowel on one side has an odd digit on the other side"—because the rule only applies to cards with a vowel on one side.

So, whatever letter is on the other side of the 3 card, the rule holds. Therefore, we don't need to check what it is.

We do, however, need to check what is on the other side of the E card, since we need to be sure it is an odd digit. And we need to check what is on the other side of the 6 card, since it could be a vowel, which would invalidate the rule that every card with a vowel on one side has an odd digit on the other.

This is one of the most famous puzzles in the psychology of reasoning, and is popularly known as the "Wason selection task" after British cognitive psychologist Peter Wason, who devised it in the 1960s. He found that fewer than 10 percent of people got the right answer, a result later replicated in other studies.

During Wason's long career at University College London, he coined the phrase "confirmation bias"—our tendency to instantly favor statements that seem to confirm our own personal beliefs, whether they are reliably true or not. Wason suggested that confirmation bias plays a role in people's failure to get the correct answer in the selection task, since what they are doing is looking for cards that confirm the rule (having a vowel and an odd digit) rather than looking for one that disproves it (having a vowel and an even digit). This puzzle also shares a similarity with "Wandering Eyes" on page 13, in that we are being asked to reason from an unknown value (in this case, the card sides that are face down) rather than a known one (the card sides that are face up). It is harder to reason from facts we *don't know* than from facts we do.

One curious aspect of the Wason selection task is that when the puzzle is phrased slightly differently, using a familiar social context, most people get the right answer.

The following four cards have a drink on one side and an age on the other. Imagine they each represent a person of that age drinking that tipple.

Whiskey	Orange Juice	25 years old	13 years old

Which of the cards do you need to turn over to check that every person drinking alcohol is over 21?

We need to turn over the Whiskey card, to make sure the age on the other side is over 21. It is also very clear that we need to turn over the "13 years old" card, to make sure they aren't drinking alcohol. We don't need to turn over the

"25 years old" card, because we know that 25-year-olds are adults and can drink whatever they like.

Rather remarkably, by introducing a social rule, the puzzle ceases to fool us.

Slippery Latitudes

List the following cities in order from north to south.

Algiers, Algeria
Halifax, Canada
Kyiv, Ukraine
Paris, France
Tokyo, Japan

Comparing latitudes between different continents is tricky, especially since our judgements tend to be based on local temperatures—but temperatures are based on many other factors than just proximity to the poles. For example, we tend to guess that, compared to North America, Western Europe is further south than it is, because the Gulf stream makes western Europe relatively warmer. And even though we think of central Japan as snowy in winter, it is at the same latitude as Crete!

The correct order, from north to south, is as follows:

Kyiv (50.5°)
Paris (48.9°)
Halifax (44.9°)
Algiers (36.8°)
Tokyo (35.7°)

A degree is about 70 miles.

The art of knowing is knowing what to ignore.

RUMI

Green Day

Anita and Bhav both work in a shop on Saturdays and Sundays. The shop provides them with two branded uniforms each—a red one and a blue one. The shop doesn't mind which color they wear to work on any one day, but it has a rule: An employee cannot wear the same color two days in a row.

Since there are only two choices, the chance of Anita and Bhav showing up on a weekend in the same color is one out of two, or 50 percent. Either they show up on the Saturday with matching colors, which means they also show up on the Sunday with matching colors, or they show up wearing different colors on the Saturday, which means they will also be wearing different colors on the Sunday.

One day the shop gives Anita and Bhav a third uniform each—a green one. The rule forbidding the same color on two consecutive days still holds.

Now what are the chances of Anita and Bhav showing up in matching uniforms on at least one day over a weekend?

a) 20 percent
b) 33 percent
c) 40 percent
d) 50 percent

d) 50 percent.

Adding an extra shirt makes no difference!

Let's say Anita wears the red uniform on a Saturday, and a blue one on the Sunday. The six Saturday/Sunday color options for Bhav are:

Anita	**Bhav**
red/blue	blue/green
red/blue	blue/red
red/blue	**red/blue**
red/blue	**red/green**
red/blue	green/red
red/blue	**green/blue**

In 50 percent of cases, the ones bolded above, there is a day with a color match. This holds for whatever combination of uniform colors Anita chooses to wear that weekend.

So there is always a 50 percent chance of a matching-shirt day.

Bemusing Basket

A large basket contains apples that are either red or green.

60 of the apples are red.
40 percent of the apples are red.

How many green apples are in the basket?

There are 90 green apples in the basket.

The calculation is not hard, but it is easy to make a mistake by trying to solve it too quickly.

If 60 apples make up 40 percent of the total number, then (dividing by four) 15 apples make up 10 percent. So we can deduce that there are 150 apples in total.

If 60 apples are red, the remaining 90 are green.

Cube Query

A piece of string is attached to one of the corners of a cube. If you pick up the string so that the cube is hanging in the air from its corner, and then lower the cube into water so that exactly half of it is submerged, what shape is the cross-section of the cube at the surface of the water?

a) Triangle
b) Square
c) Hexagon
d) Another shape

c) Hexagon.

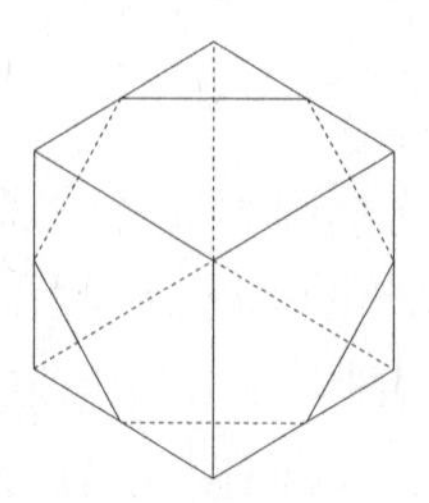

I have asked this question many times to audiences of·children and adults, and almost no one gets it right. Our brains are incredibly bad at mentally picturing in three dimensions. The cube is a very familiar shape, probably the most basic three-dimensional object after the sphere, yet its geometrical properties are surprising.

Unless you have unbelievable powers of visualization, or are a sculptor or architect, the way to solve this problem is to deduce the correct answer. So . . .

A cube has six sides. If it is hanging from a corner, such that half of its volume is under the water, then the cross-section must be halfway between opposite corners. The cross-section, therefore, must cut through all six faces: The two-dimensional shape that results must be a symmetrical six-sided shape—i.e., a regular hexagon.

Random Family

In a certain city, children are born male or female, and the chances of a child being born a boy or a girl are equal. A third of families in this city have two children, a third have one child, and a third have no children.

If a family is chosen at random, and this family has at least one child, the chance that it has a second child is 50 percent, because either the family is one of those with two children, or it is one of those with a single child—and both options are equally likely.

Now let's say a family is chosen at random, and this family has at least one girl.

What is the chance that *this* family has another child?

a) 40 percent
b) 50 percent
c) 60 percent
d) 66 percent

The answer is c) 60 percent.

Remarkably, although you wouldn't think it, a mention of one child's gender makes it 10 percent more likely that a family is a two-child one. It is very strange to think that specifying the gender makes such a big difference to the number of children, since all children necessarily have a gender, and all are equally likely to be either a boy or a girl.

But here's how the math works out:

For every 12 randomly chosen families, there are 12 equally likely combinations of children:

no kids	G	GG
no kids	G	GB
no kids	B	BG
no kids	B	BB

As we can see, a third have no children, a third have one, and a third have two. The chances of a boy or a girl are equal, and in the final column the kids are listed in birth order.

Five families in 12 have at least one girl (G, G, GG, GB, BG), and in three of them (GG, GB, BG) the family has another child. Thus the chance a family with a girl has another child is 3/5, or 60 percent.

Flummoxing Fly

A fly has landed on the surface of a glass bowl which is sitting on a very sensitive digital scale. All of a sudden, the fly takes off.

What happens to the readout on the scale?

a) It goes up.
b) It goes down.

Both!

Or to be more specific, a) then b). The reading on the scale goes up, and then it goes down.

Why?

Imagine you are standing on your bathroom scale. To propel yourself to jump off, you push down on the scale, which momentarily registers a heavier weight.

The situation with the fly launching itself from the bowl is similar. As it takes flight, the fly's wings push the air down, which, in turn, pushes the bottom of the bowl, and the sensitive scale registers an increase in weight. But when the fly is high enough so the air movement does not reach the surface, the measured weight will return to that of the empty glass bowl.

Double Bad

Here's a terrible game in which you are guaranteed to lose all your money:

Bad Game A: Each time you play you lose $1.

Here's another terrible game, in which you are also guaranteed to go bust:

Bad Game B: Each time you play, you count your money. If you have an even number of dollars, you win $3. If you have an odd number of dollars, you lose $5.

Given that both these games are designed to make you penniless, is there a way to play them together that guarantees you actually *make* money?

Yes, there is.

If you play both games alternately, not only will you not go bust, you will actually make money.

Let's say you have $10. And let's say you start with **Bad Game B**, followed by a go at **Bad Game A**.

Here are your totals after each game:

B: $13
A: $12
B: $15
A: $14

Alternatively, let's say your starting sum is $9. And this time you begin with a round of **Bad Game A**:

A: $8
B: $11
A: $10
B: $13

In both scenarios, your money piles up.

This puzzle is an example of Parrondo's paradox, a counterintuitive result from game theory named after the Spanish physicist Juan Parrondo, who formulated it in 1996. According to the paradox, it's possible to create two games with a higher probability of losing, and then develop a winning strategy by playing them alternately.

A very little key will open a very heavy door.

CHARLES DICKENS

In-Law Enigma

Sisters and brothers, I've none at all. But my dear old mum is your mum's mum-in-law.

Who am I?

I'm your father.

There's nothing counterintuitive about this riddle, but it is hard to keep all the family relationships in your head at the same time, so it's all too easy to slip up or overthink things.

Möbius Mind-Mangler

Here are two Möbius strips stuck together at right angles. Each strip has a line drawn down the middle.

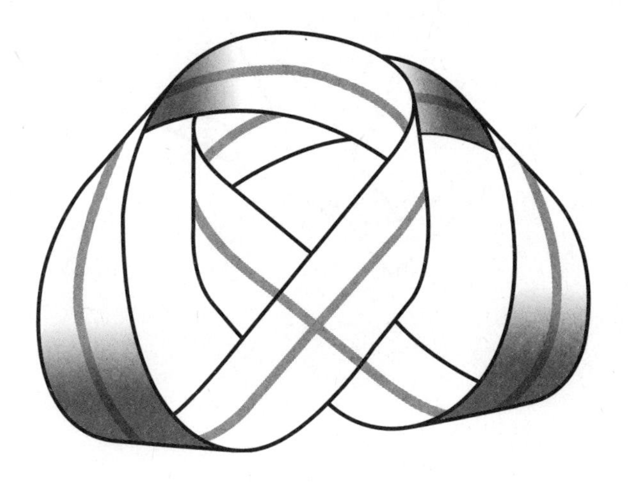

What do you get when you cut along both lines?

Two interlocking hearts. It is genuinely surprising—and delightful!

Hospital Head-Scratcher

A certain town is served by two hospitals of different sizes. In the larger hospital about 45 babies are born each day, and in the smaller hospital about 15 babies are born each day. As you know, about 50 percent of all babies are boys. However, the exact percentage varies from day to day. Sometimes it may be higher than 50 percent, sometimes lower.

For a period of one year, each of the two hospitals recorded the days on which more than 60 percent of newborn babies were boys.

Which hospital do you think recorded more of these days?

a) The larger hospital
b) The smaller hospital
c) They are about the same (that is, within 5 percent of each other).

The correct answer is b) the smaller hospital.

This puzzle requires knowledge of what is known in statistics and probability theory as the "law of large numbers": That is, the more examples of a random result you have, the closer the results will be to the expected average. While this law might seem obvious when you think about it, most people don't take it into account.

When there are only a few examples, it is very possible that the number of newborn boys will diverge widely from the expected average of 50 percent. There's a 1 in 32 chance of getting five boys in a row, for example, which would skew the results of the small hospital more than it would those of the large one. Over time, and with more data, the results will match the 50/50 ratio of boys to girls much more closely.

Again, this puzzle originates from the work of Daniel Kahneman and Amos Tversky, who spent their careers analyzing psychological biases. In the 1972 paper in which they looked at this question, they found that 56 percent of respondents answered c)—that the two hospitals recorded about the same number of days on which more than 60 percent of births were boys.

You should never be ashamed to admit that you are wrong. It only proves that you are wiser today than yesterday.

JONATHAN SWIFT

Double Trouble

A and B are 10 miles apart. I cycle from A to B at 10 mph.

How fast must I cycle back from B to A in order to double my average speed for the whole journey?

a) 30 mph
b) Somewhere between 30 mph and 300 mph
c) Faster than 300 mph

If there is a solution to this question, the answer is c).

I would actually have to cycle the 10 miles back from B to A at a speed of infinity mph.

This question is two puzzles wrapped in one. That is, there is an obvious wrong answer—which is a) 30 mph. And there is a less obvious wrong answer—b) Somewhere between 30 mph and 300 mph—which is the one I was hoping you would choose.

First, let's dismiss answer a) 30 mph.

If you cycled from A to B at 10 mph, and from B to A at 30 mph, then it is certainly true that, at every point traveled, the speed would be 20 mph on average. You would be passing through each point at both 10 mph and 30 mph.

However, the question asked about the average speed *overall*, not the average speed at particular points.

Speed is the distance traveled divided by time. So, if you travel 10 miles at 10 mph and 10 miles at 30 mph, the time spent at the higher speed is less, since you get there faster. So it would seem logical that I would need to return from B to A at a higher speed than 30 mph in order for the average speed to double. I didn't want you to do the calculations, so I gave you the option of between 30 mph and 300 mph.

In fact, you didn't need to do any calculations at all to realize that the question is ridiculous. For an average speed of 20 mph, I would have to cycle the entire round trip (20 miles) in one hour. Yet it has already taken me one hour to get to B, just halfway! In order to double the speed overall, I would need to travel instantaneously (at infinity mph?) back to A.

Looking for people born on
NOVEMBER
22

It's My Party

How many people do you have to meet until it is more likely than not that you have met someone who shares your birthday?

a) 73
b) 183
c) 253
d) 365

The answer is c) 253 people, which to many seems a counterintuitively large amount.

If you could guarantee that you'd only meet people who each had a different birthday, then you would need to meet 183 people to have a 50/50 chance of meeting someone with your birthday. (183 accounts for half of the days in the year.)

However, if you are meeting people at random, there will be many shared birthdays among those you meet (see "Jolly Good Fellows" on page 113), and these doubled-up birthdays mean that your search takes considerably longer. You will have to socialize a lot more than you might have expected!

Weighty Question

When I put 100 kg of melons out in the garden this morning, 99 percent of their weight was water. Now, after a day in the blazing sun, only 98 percent of their weight is water.

How much do the melons weigh now?

a) 50 kg
b) 98 kg
c) 99 kg

a) The melons weigh 50 kg.

Most people say 98 kg or 99 kg, since it feels like a 1 percent drop in water weight should be reflected by a comparable percentage drop in total weight. In fact, the relevant detail is the percentage of weight that is *not* water, which doubles.

To understand this, an illustration is helpful. The grid on the left shows the ratio in the melons of non-water (the shaded cell) to water (the white cells) in the morning. And the grid on the right shows the same ratio in the evening.

Morning
99 percent water:
total 100 kg

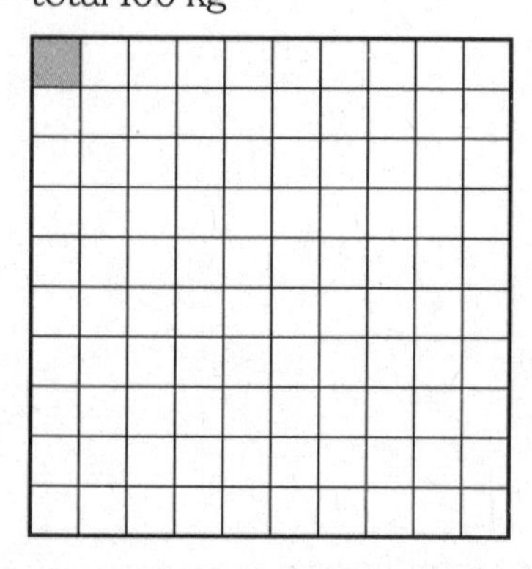

Evening
98 percent water:
total 50 kg

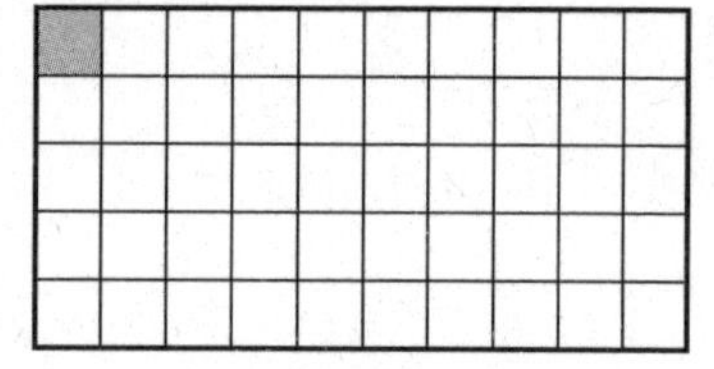

In the morning, we can readily calculate that there are 99 kg of water and 1 kg of non-water.

By evening, that 1 kg accounts for 2 percent of the total weight (up from 1 percent)—so the new total must be 50 kg.

From a drop of water a logician could infer the possibility of an Atlantic or a Niagara without having seen or heard of one or the other.

SIR ARTHUR CONAN DOYLE

Curve-ball

In the image below, the curved grey line is made up of a semicircle joining A to B, and a semicircle joining B to C. A black semicircle joins A to C.

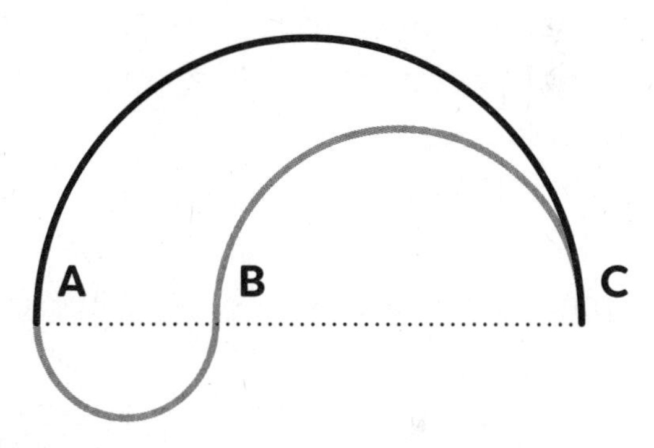

Which is true?

a) The grey curved line is longer.
b) The black curved line is longer.
c) The grey and black curved lines have the same length.
d) There is not enough information to determine the lines' relative lengths.

c) The curved lines have the same length.

According to Presh Talwalkar of the popular YouTube channel "Mind Your Decisions," this puzzle—a question adapted from the SAT—is one that "everyone" gets wrong.

To solve the puzzle, we need to know that the circumference of a circle is π (pi) times the diameter. Let's assume that the length AC is x and that AB is y (so BC is $x - y$).

The length of the black semicircle is: $\frac{\pi x}{2}$

The length of the smaller grey semicircle (A to B) is: $\frac{\pi y}{2}$

The length of the larger grey semicircle (B to C) is: $\frac{\pi(x-y)}{2} = \frac{\pi x}{2} - \frac{\pi y}{2}$

The combined length of the grey semicircles is therefore:

$$\frac{\pi y}{2} + \frac{\pi x}{2} - \frac{\pi y}{2} = \frac{\pi x}{2}$$

Thus, both the black and grey curved lines have equal length. Note that the proof did not require any named lengths (apart from the variables x and y). In other words, whatever the relative positions of A, B, and C, the black and grey curved lines have the same length.

When you doubt your power, go give power to your doubt.

HONORE DE BALZAC

Chess Mates

Alyssa and Boyan spent the weekend playing chess against the computer.

On Saturday, Alyssa won a greater percentage of games than Boyan.

On Sunday, Alyssa won a greater percentage of games than Boyan.

Here's a table of the games they played and won:

	Alyssa			Boyan		
	Games played	Games won	Win percentage	Games played	Games won	Win percentage
Saturday	2	2	100 percent	8	7	87.5 percent
Sunday	10	7	70 percent	2	1	50 percent

Over the weekend, who won a greater percentage of games?

a) Alyssa

b) Boyan

b) Boyan won the greater percentage of games played over the weekend.

Since Alyssa did better than Boyan on Saturday *and* Sunday, you might have expected her to have done better over both days. In fact, when you combine the data, the opposite is the case. Boyan's win percentage is higher overall. We can tot up the tallies from the table. Alyssa won nine out of 12 games, giving a win percentage of 75 percent. But Boyan won eight out of 10 games, giving him a win percentage of 80 percent.

The curious phenomenon whereby, sometimes, a result seen across individual sets of data is reversed when all the data is combined is called "Simpson's paradox," after the British statistician Edward H. Simpson (1922–2019), who described it in 1946. (In his early twenties, during the Second World War, he was working as a codebreaker at Bletchley Park).

Simpson's paradox is not a logical paradox, meaning that there is nothing self-contradictory about it, but it is incredibly counterintuitive. The paradox appears widely in medicine and social science and is often used to show how statistics, when misused or interpreted incorrectly, can be misleading. When aggregating data, one needs to be very careful!

One of the best-known cases of Simpson's paradox concerns gender bias at the University of California, Berkeley. In 1973, the university admitted 44 percent of men applying for graduate studies, and 35 percent of women. The university officials, worried about being sued for discrimination, looked more closely at the data and found that, in fact, when examined on a department-by-department basis, there was a slight bias in favor of women! The reason for the discrepancy was that women were applying in greater numbers to much more competitive departments, like English, whereas men were applying to less competitive departments where it was easier to get in, like engineering. Depending on how you looked at the data, you could argue that it was biased toward women, or against them.

In a medical context, Simpson's paradox can crop up when different treatments are compared.

For example, let's say that treatment A is tested on a group of people, and is shown to be more successful than treatment B, which is tested on another group of people. This is like saying that Alyssa won more chess matches than Boyan on Saturday. Now let's say we test treatment A against a third group of people, and B against a fourth group, and again A is more successful than B. This is like saying that Alyssa won more chess matches than Boyan on Sunday. However, although treatment A did better than treatment B in both comparisons, it may be the case that, overall, treatment B had better results than treatment A.

Brain-Stretcher

An ant at one end of a 1 km elastic rope starts to walk to the other end at a rate of 1 cm per second. At the end of each second, however, the rope is stretched by a kilometer. So, after a minute the rope is 60 km long, and after an hour it is 3,600 km long.

Will the ant ever reach the other end?

a) Yes
b) No

(For the purposes of this puzzle, assume that the ant does not age and never tires as it walks, and that the rope will carry on stretching forever.)

The answer is a) Yes, the ant eventually reaches the end of the rope.

Of course it does! It seems so wildly unlikely that the slow ant will conquer the speedily expanding rope that—according to the logic of this book—you know it *must* do. The hard bit is to work out how or rather *why*.

The proof, which I will get to below, is the most mathematical in this book, and relies on a theorem that you may have learned (or forgotten) from school. But you can also catch the gist of why the ant gets there by considering its progress. After one second, the ant is 1 cm along the rope. But this immediately becomes 2 cm, since when a 1 km rope is stretched by 1 km, the distance between any two points is doubled (i.e., a distance of 1 cm becomes 2 cm). After two seconds, the ant is 3 cm along, which immediately becomes 4.5 cm when the rope is stretched by another 1 km, because when a 2 km rope is stretched by 1 km, the distance between any two points is increased by 50 percent (i.e., a distance of 3 cm increases by 1.5 cm, making 4.5 cm). Gradually the critter is inching along, given a helping hand by the stretching rope.

Here's the mathy bit. Since 1 cm is 1/100,000 of a kilometer, we can write a formula for the fraction of the rope traveled by the ant after n seconds:

$$\frac{1}{100{,}000} \times \left(1 + \frac{1}{2} + \frac{1}{3} + \frac{1}{4} + \ldots + \frac{1}{n}\right)$$

The sequence in parentheses is known as the "harmonic series." When you meet the harmonic series in school, a simple proof shows that it is "divergent," meaning that it gets bigger and bigger all the way to infinity. In other words, the series will exceed any finite number if you keep adding enough terms. The series will therefore exceed 100,000 for some value of n and, when it does, the ant will fall off the other end of the rope.

USA

Transatlantic Trip

Which of the following US states is closest to Africa?

a) Florida
b) Maine
c) Massachusetts
d) North Carolina

a) Maine is closest to Africa, followed by Massachusetts, North Carolina, and Florida.

Our understanding of the relative positions of countries is primarily down to maps, which are usually two-dimensional, and which distort distances as you move toward the poles. The east coast of North America leans toward Europe and Africa much more than you might think from reading a standard map.

Puzzling Pay

You are offered a job with a starting salary of $50,000 per annum. You are asked to choose between two packages:

Plan A: Every year, your annual salary increases by $4,000.

Plan B: Every half year, your half-yearly salary increases by $1,000. (Meaning that you take home $1,000 more every six months.)

Which is the most lucrative plan?

Plan A sounds much better, but a quick analysis reveals that Plan B is the one to choose.

If you opted for Plan A, in your first year you would earn $50,000. In your second year you would earn $54,000, and so on.

If you chose to go for Plan B, in your first half year you would earn $25,000. Then you would get a raise, and in the next half year you would earn $26,000. So the total for year one is $51,000. In the second year you would earn $27,000 in the first six months and $28,000 in the second half of the year, making your annual salary $55,000 in total.

As more years accumulate, Plan B will put you nicely ahead in the earning stakes.

Aquarium Tragedy

Fatima had forty fish. All but thirty-nine died.

How many fish were left?

Thirty-nine of Fatima's fish survived.

Eleven vs. Ten

When you and your pal go out together for dinner, you always flip a coin to decide who pays. Heads they pay, and tails you pay. In this way, 50 percent of the time they pay, and 50 percent of the time you will foot the bill. It's a fair system.

One day your pal suggests a different scheme. They say they will flip the coin 11 times, while you will flip it only 10 times. Under the new rules, the person who flips heads the most times will pay for dinner. (But if you both flip the same number of heads, *you* will pay.) Your pal has more attempts than you, so on average will flip more heads than you do.

Are these new rules in your interest, or not?

Surprisingly, even though your pal has more flips of the coin, the chance of either of you paying remains at 50/50.

Compare your pal's first 10 flips and your 10 flips. If your friend flips more heads, they pay the bill, whatever the outcome of their 11th coin. If they flip fewer heads, you pay, whatever the outcome of their 11th coin, because after their final flip either there will be equal numbers of heads (you pay), or you'll still have more heads (you pay). Since you have both flipped exactly 10 coins, the chances of one of you having flipped more heads than the other is 50/50. So we need to look at what happens when you both have the same number of heads after 10 flips.

In this case, your friend's final flip is a tie-breaker—heads they pay; tails you pay.

Since this result is again 50/50, the new scheme offers a 50/50 chance of either of you treating the other one to dinner.

Falling Target

An archer is aiming a crossbow directly at an apple in a tree. But the instant the archer shoots, the apple falls from the tree.

Assuming the archer shoots with enough force for the bolt to cover at least the distance to the tree, and ignoring air resistance, does the archer's bolt hit the apple?

a) Yes
b) No

a) Yes, the archer's bolt hits the apple.

The apple is falling thanks to gravity, but so is the bolt! The bolt's vertical drop will be the same as that of the apple.

So if the crossbow was pointed directly at the apple at the moment of shooting, the bolt will reach the apple as it falls, whatever its initial velocity.

Sources

Puzzles without sources are classics I have refashioned. If I have inadvertently missed anyone, please let me know.

Sources are listed by page number.

11. Richard Wiseman, *101 Bets You Will Always Win*, 2016

13. Hector Levesque, from "Rational and Irrational Thought: The Thinking That IQ Tests Miss," in *Scientific American*, 2009; James Grime, youtube.com/singingbanana

21. A. J. Friedland, *Puzzles in Math & Logic*, 1970. Quoted on stanwagon.com.

23. A. J. Friedland, *Puzzles in Math & Logic*, 1970. Quoted on stanwagon.com.

25. Thomas Povey, *Professor Povey's Perplexing Problems*, 2015

29. robeastaway.com

35. Daniel Kahneman and Amos Tversky, "Availability: A Heuristic for Judging Frequency and Probability," in *Cognitive Psychology*, 1973

43. Peter Winkler

45. Steven E. Landsburg, *Can You Outsmart an Economist?*, 2018

47. Des MacHale

49. Rob Eastaway and Brian Hobbs, *Headscratchers*, 2023

57. Adam Rutherford, waterstones.com/blog/family-fortunes-adam-rutherford-on-how-were-all-related-to-royalty

63. Ivan Moscovich, *The Big Book of Brain Games*, 2006

69. Steven E. Landsburg, *Can You Outsmart an Economist?*, 2018

83. Andrew Meyer and Shane Frederick, "The formation and revision of intuitions," *Cognition*, 2023

119. stanwagon.com

133. Henk Tijms, *Basic Probability: What Every Math Student Should Know*, 2019

143. P. C. Wason, "Reasoning about a rule," *Quarterly Journal of Experimental Psychology*, 1968

163. Alex Lvovsky

165. Wikipedia: Parrondo's paradox

173. Daniel Kahneman and Amos Tversky, "Subjective probability: A judgement of representativeness," *Cognitive Psychology*, 1972

187. youtube.com/MindYourDecisions

183. Peter Winkler, *Mathematical Puzzles: A Connoisseur's Collection*, 2017

205. Peter Winkler, *Mathematical Puzzles*, 2021

207. Ivan Moscovich, *The Big Book of Brain Games*, 2006

Image Credits

Images are listed by page number.

89. Illusion created using a photograph by Irina Bg, Russia. Original © Irina Bg/Shutterstock.

91, 92. The "Bemusing Balls" illusion was originally titled the "Confetti illusion." Both images are © David G. Novick, the University of Texas at El Paso.

93. "Card Sharp" puzzle image is from a mobile game called "Stenciletto," created by the Smiley World Games team, inspired by Grace Arthur's stencil design test. Artwork provided by Jane Braybrook © stenciletto.com.

95. "Pavement Poser" image is a photograph of ephemeral public artwork created by artist Michael Neff from his "Night Shadows" project. Photograph and artwork © Michael Neff.

99. "Murky Med" map illustration, originally titled "Watercolor illusion," © Baingio Pinna.

101. "Café Confusion" artwork © Victoria Skye.

103. "Fool House" playing cards © Mannaggia/ stock.adobe.com.

158. "Cube Query" cross-section © Edmund Harriss.

Acknowledgments

Thank you to the many people who have shared puzzles and ideas with me over the years, in particular Gary Antonick, Jane Braybrook, Rob Eastaway, James Grime, Tim Harford, Tanya Khovanova, Alex Lvovsky, Des MacHale, Adam Rubin, Victoria Skye, Chris Smith, and Peter Winkler. The readers of my *Guardian* column are also an invaluable source of brainteasers. Keep them coming please.

At The Experiment, my US editor, Karen Giangreco, brilliantly solved all the publishing puzzles my manuscript threw at her, as did Jenny Dean at Square Peg in the UK. She was ably helped by assistant editor Emily Martin and managing editor Graeme Hall. Copy editor Rosemary Davies made me think at least twice per sentence and proofreader Ben Sumner got the things right that (almost) everyone gets wrong. Thanks to publisher Marianne Tatepo for being so enthusiastic about the book in the first place.

I'm grateful to Arnaud Boutin for his stylish illustrations and his ability to capture my unruly hair. A duo of dextrous Dans did the design. Dan Mogford put together the cover, and Dan Prescott of Couper Street Design composed the interior pages and the technical drawings. Thanks also to the book's publicist Mia Quibell-Smith, and Amelia Rushen and Gabriela Quattromini in marketing. Konrad Kirkham was in control of production, which nudges up the number of words in this book beginning with K. (See page 35.)

My agent Claire Conrad I thank twice: first, for listening to me when I suggested the idea as we were crossing a road in central London, and second, for not getting us run over.

As you know by now, I like to get simple puzzles wrong. But more than anything, I like to think thrice: of Natalie and our boys Zak and Barnaby, who have always been here for love and support during the writing of this book.

About the Author

Alex Bellos holds a degree in mathematics and philosophy from Oxford University. His bestselling books *Here's Looking at Euclid* and *The Grapes of Math* have been translated into more than twenty languages and were both short-listed for the Royal Society Science Book Prize. His puzzle books include *The Language Lover's Puzzle Book*, *Can You Solve My Problems?*, *Puzzle Ninja*, and *Perilous Problems for Puzzle Lovers*, and he is also the coauthor of the mathematical coloring books *Patterns of the Universe* and *Visions of the Universe*. He has launched an elliptical pool table, LOOP. He writes a popular-math blog and a puzzle blog for *The Guardian*, and he won the Association of British Science Writers award for best science blog in 2016. He lives in London.

alexbellos.com | @alexbellos